D1574506

Iain T. Adamson

# A Set Theory Workbook

Birkhäuser
Boston • Basel • Berlin

Iain T. Adamson
Department of Mathematics
The University of Dundee
Dundee DD1 4HN
Scotland

**Library of Congress Cataloging In-Publication Data**

Adamson, Iain T.
A set theory workbook / Iain T. Adamson.
p. cm.
Includes index.
ISBN 0-8176-4028-2 (acid-free paper)
1. Set theory. I. Title.
QA248.A27 1997 97-36639
511.3'22--dc21 CIP

AMS Subject Classifications: 04-01, 03-01

Printed on acid-free paper


*Birkhäuser* 

ISBN 0-8176-4028-2
ISBN 3-7643-4028-2

Typeset by the author in LaTeX.
Printed and bound by Quinn-Woodbine, Woodbine, NJ
Printed in the U.S.A.

9 8 7 6 5 4 3 2 1

# Contents

# PREFACE

This book is a companion to *A general topology workbook* published by Birkhäuser last year. In an ideal world the order of publication would have been reversed, for the notation and some of the results of the present book are used in the topology book and on the other hand (the reader may be assured) no topology is used here.

Both books share the word *Workbook* in their titles. They are based on the principle that for at least some branches of mathematics a good way for a student to learn is to be presented with a clear statement of the definitions of the terms with which the subject is concerned and then to be faced with a collection of problems involving the terms just defined. In adopting this approach with my Dundee students of set theory and general topology I found it best not to differentiate too precisely between simple illustrative examples, easy exercises and results which in conventional textbooks would be labelled as Theorems. (I have used the title *Theorem* for some exercises occasionally to label them with their traditional names but mainly to indicate not so much that they are difficult but that they are important and useful.) And, although readers may not find the hints I have provided in this book very generous, the students in my classes got none at all. Not only that, they were not shown the solutions to the exercises until they had made serious attempts to solve them for themselves. That was easy to arrange in the classroom situation but readers of the book can cheat by looking at the full answers provided in Part II. I should like to encourage them not to do that—and to remind them that this is a *Workbook* and that they will derive more benefit from it if they work seriously at the exercises before turning to the solutions.

The titles of Chapters 2 to 13 show that the contents of the book are entirely conventional for an advanced undergraduate or beginning graduate course in set theory. The treatment is intended to be relatively informal, in the sense that definitions and proofs are given for the most part in (hopefully clear) standard "mathematical English" not in logical symbolism. I hope that the reference in the Introduction to "the first order predicate calculus with equality" will not be found off-putting—it is, after all, only a shorthand way of describing the logical framework that we use as soon as we start to do mathematics seriously.

This is the fourth book in which I have had the help and advice of Nick Dawes with my Latex typesetting; as he breathes a sigh of relief that this is (I think) the last, I want to put on record my gratitude to him for his calm and calming expert assistance. I must again thank the several generations of students in The University of Dundee who wrestled successfully with the material presented here and whose success with it suggested that I turn it into this book. I am very grateful also to my friends—and particularly my wife and daughter—who encouraged me to complete it.

Iain T. Adamson

Dundee, Scotland
September 1997

# INTRODUCTION

# Introduction

Cantor's work on the foundations of set theory started about 1875. For about twenty years it met with distrust and hostility from mathematicians and indifference from philosophers. In the early 1890's set theory became fashionable and was widely applied in analysis and geometry.

In 1895 Cantor discovered a paradox, not then published but communicated to Hilbert in 1896; it was rediscovered by Burali-Forti in 1897 and is called *Burali-Forti's paradox of the greatest ordinal number.* (We shall describe the paradox in Chapter 8.) Cantor and Burali-Forti could not resolve this paradox, but it was not taken too seriously, partly because it was in a rather technical region.

*Russell's paradox* (1902) shook the foundation of work by Cantor, Frege and Dedekind. (Let $R$ be the set of all sets which are not members of themselves. Then $(R \in R) \Longrightarrow (R \notin R)$, which is a contradiction; but $(R \notin R) \Longrightarrow (R \in R)$, which is again a contradiction.)

Paradoxes can be avoided in most "everyday" regions of applications of set theory: people wanted to believe in set theory. Axiomatic set theory is an attempt to satisfy their needs without involving paradoxes. The paradoxes appear to be connected with collections (such as the collection of all sets) which are in some sense "too big", and the two main approaches to developing formal theories which avoid the paradoxes deal with such collections in different ways: roughly speaking the *Zermelo-Fraenkel* theory lays down strict criteria for the admissibility of collections, so that collections which are too large are not admitted; the *Von Neumann-Bernays-Gödel* theory admits the large collections but attempts to avoid the paradoxes by not allowing them to be members of collections.

We follow here Kelley's modification of the Von Neumann-Bernays-Gödel approach.

Like every other axiomatic theory, axiomatic set theory is based on an underlying system of logic. For the basic logical discipline of axiomatic set theory we take the **first-order predicate calculus with equality**. In this calculus we have

(1) a (potentially) infinite list of **variables**, $x$, $y$, $z$, $w$, $\ldots$, $x'$, $y'$, $z'$, $w'$, $\ldots$ (these are the basic terms of the theory);

(2) **connectives** expressing

**negation** ($\neg$), **conjunction** ($\wedge$), **disjunction** ($\vee$), **implication** ($\Longrightarrow$) and **equivalence** ($\Longleftrightarrow$)

($\neg P$ is read "not $P$", $P \wedge Q$ as "$P$ and $Q$", $P \vee Q$ as "$P$ or $Q$", $P \Longrightarrow Q$ as "$P$ implies $Q$ or "if $P$ then $Q$"and $P \Longleftrightarrow Q$ as "$P$ is equivalent to $Q$" or "$P$ if and only if $Q$" where $P$ and $Q$ are sentences of the theory);

(3) **universal** and **existential quantifier** symbols ($\forall$ and $\exists$) expressing "for all" and "there exists";

(4) the **equality** sign ($=$);

(5) auxiliary symbols such as commas, parentheses, brackets.

These symbols are subject to the usual conditions familiar from informal studies of logic—for example,

$$(\forall A)(\forall B)(\neg(A \wedge B) \Longleftrightarrow ((\neg A) \vee (\neg B))),$$
$$(\forall A)(\forall B)(\neg(A \vee B) \Longleftrightarrow ((\neg A) \wedge (\neg B))),$$
$$(\forall P)(\neg(\forall x)P \Longleftrightarrow (\exists x)(\neg P)),$$
$$(\forall P)(\neg(\exists x)P \Longleftrightarrow (\forall x)(\neg P)) \text{ and}$$
$$(\forall P)(\neg(\neg P) \Longleftrightarrow P).$$

A variable $x$ in a sentence $P$ (often written $P(x)$ if we wish to emphasize that $x$ occurs in $P$) is said to be **bound** if it is preceded in $P$ by either $\forall x$ or $\exists x$; otherwise $x$ is **free**. A sentence is **closed** if it contains no free variables; otherwise it is **open**. An open sentence which involves $x$ as a free variable is called a **condition on** $x$.

The equality symbol is subjected to the following conditions:

(1) $(\forall x)(x = x)$;

(2) $(\forall x)(\forall y)((x = y) \Longrightarrow (y = x))$;

(3) $(\forall x)(\forall y)(\forall z)(((x = y) \wedge (y = z)) \Longrightarrow (x = z))$;

(4) for every $x$ and $y$ and every open sentence $P(x)$ in which $y$ does not occur as a bound variable, if $x = y$ then $P(x) \Longleftrightarrow P(y)$.

(In (4), $P(y)$ denotes the sentence obtained on replacing every occurrence of $x$ in $P(x)$ by $y$.)

# Part I

# EXERCISES

# Chapter 1

# FIRST AXIOMS OF THE THEORY NBG

This theory is obtained by adjoining to the first-order predicate calculus with equality a new binary predicate $\in$. The "atomic sentences" of NBG are formulæ of the forms $x = y$ and $x \in y$.

The sentence $x \in y$ is read "$x$ **is a member of** $y$", "$x$ **belongs to** $y$", "$x$ **is contained in** $y$", "$x$ **is an element of** $y$", "$y$ **contains** $x$". The terms in the theory NBG are called **classes**; a class $x$ is said to be a **set** if there exists a class $y$ such that $x \in y$ (so a set is a class which is a member of another class); classes which are not sets are called **proper classes**. This is actually a very unfortunate terminology—it would have been more intuitive to call such classes *improper*.

The axioms of any theory are the basic statements from which all other statements (theorems) of the theory are deduced. We introduce the axioms of NBG one by one throughout the book. The first axiom gives us the means of proving that two classes are equal.

**Axiom 1 (Axiom of Extensionality).** For all $x$ and $y$ we have $x = y$ if and only if, for all $z$, $z \in x$ is equivalent to $z \in y$.

(So two classes are equal if and only if every member of each is also a member of the other.)

The second axiom, more properly called an *axiom scheme* since it gives us a method of producing one axiom for each open sentence, allows us to introduce new terms of the theory, i.e. new classes, one for each open sentence.

**Axiom 2 (Axiom Scheme of Classification).** For each open sentence $P(x)$ there exists a class which consists precisely of those sets which satisfy the condition $P(x)$.

The class whose existence is postulated by Axiom 2 is denoted by $\{x : P(x)\}$; thus $\{x : P(x)\}$ is a term of the theory NBG and the assertion $u \in \{x : P(x)\}$ is true if and only if $u$ is a set and $P(u)$ is true.

The **union** and **intersection** of two classes are defined in exactly the same way as the union and intersection of two sets in naïve set theory: if $A$ and $B$ are classes, then

$$\begin{aligned} A \cup B &= \{x : (x \in A) \vee (x \in B)\}, \\ A \cap B &= \{x : (x \in A) \wedge (x \in B)\}. \end{aligned}$$

So a set $x$ is a member of $A \cup B$ if and only if it is a member of either $A$ or $B$ (or both); $x$ is a member of $A \cap B$ if and only if it is a member of both $A$ and $B$.

The usual properties of these unions and intersections are established as in naïve set theory. Namely, we have the properties known as **idempotence**,

$$(\forall X)(X \cup X = X), \ (\forall X)(X \cap X = X),$$

**associativity**,

$$(\forall X)(\forall Y)(\forall Z)(X \cup (Y \cup Z) = (X \cup Y) \cup Z),$$

$$(\forall X)(\forall Y)(\forall Z)(X \cap (Y \cap Z) = (X \cap Y) \cap Z),$$

**commutativity**,

$$(\forall X)(\forall Y)(X \cup Y = Y \cup X), \ (\forall X)(\forall Y)(X \cap Y = Y \cap X),$$

and **distributivity**,

$$(\forall X)(\forall Y)(\forall Z)(X \cup (Y \cap Z) = (X \cup Y) \cap (X \cup Z)),$$

$$(\forall X)(\forall Y)(\forall Z)(X \cap (Y \cup Z) = (X \cap Y) \cup (X \cap Z)).$$

We write $x \notin y$ as an abbreviation for $\neg(x \in y)$. Then for each class $A$ we define the **complement** of $A$ to be the class

$$\sim A = \{x : x \notin A\}.$$

This is a definition which is inappropriate in naïve set theory and in Zermelo-Fraenkel set theory, since the complement of a set is "very large" and so possibly productive of paradoxes.

For any two classes $A$ and $B$ we define the **difference** of $A$ and $B$, or **complement of $B$ relative to $A$**, to be the class

$$A \sim B = \{x : (x \in A) \wedge (x \notin B)\} = A \cap (\sim B).$$

For all classes $A$ we have $\sim (\sim A) = A$ and for all classes $A$ and $B$ we have the De Morgan rules $\sim (A \cup B) = (\sim A) \cap (\sim B)$ and $\sim (A \cap B) = (\sim A) \cup (\sim B)$.

Two special classes are introduced, the **null class** $\emptyset$ and the **universe** $V$:

$$\emptyset = \{x : x \neq x\} \quad \text{and} \quad V = \{x : x = x\}.$$

For each class $A$ we have $A \notin \emptyset$, $A \cup \emptyset = A$, $A \cap \emptyset = \emptyset$, $A \cup V = V$, $A \cap V = A$ and $A \in V$ if and only if $A$ is a set. Further $\sim \emptyset = V$ and $\sim V = \emptyset$.

Let $A$ be a class; the **union** and **intersection** of the class $A$ are the classes

$$\begin{aligned} \bigcup A &= \{x : (\exists y)((y \in A) \wedge (x \in y))\}, \\ \bigcap A &= \{x : (\forall y)((y \in A) \Longrightarrow (x \in y))\}. \end{aligned}$$

Thus a class $C$ belongs to $\bigcup A$ if and only if (1) $C$ is a set and (2) $C$ belongs to at least one of the members of $A$; $C$ belongs to $\bigcap A$ if and only if (1) $C$ is a set and (2) $C$ belongs to every member of $A$.

These definitions, specialised to the case where $A = \emptyset$, produce the results of Exercise 1 when we apply Axiom 1.

**Exercise 1.** Prove that $\bigcap \emptyset = V$ and $\bigcup \emptyset = \emptyset$.

If $A$ and $B$ are classes such that every member of $A$ is also a member of $B$, i.e. such that we have

$$(\forall x)((x \in A) \Longrightarrow (x \in B)),$$

we say that $A$ **is included in** $B$, $B$ **includes** $A$ or $A$ **is a subclass of** $B$ and we write $A \subseteq B$ or $B \supseteq A$. (If $A$ is a set and $A \subseteq B$ we say that $A$ is a **subset** of $B$.) If $A \subseteq B$ and there is at least one set $b$ such that $b \in B$ but $b \notin A$, we say that $A$ **is properly included in** $B$, $B$ **properly includes** $A$ or $A$ **is a proper subclass of** $B$ and we write $A \subset B$ or $B \supset A$.

(Many mathematicians read the sentence $A \subseteq B$ as "$A$ is contained in $B$" or "$B$ contains $A$" and similarly for $A \subset B$. We prefer to use "contain" only to refer to the relation between an element and a set of which it is a member: thus we read $x \in A$ as "$x$ is contained in $A$" and "$A$ contains $x$".)

For each class $A$ we have $A \subseteq A$, $\emptyset \subseteq A$ and $A \subseteq V$. The Axiom of Extensionality can be restated in the form: $A = B$ if and only if $A \subseteq B$ and $B \subseteq A$. It is an immediate consequence of the definition of inclusion that for all classes $A$, $B$, $C$ for which $A \subseteq B$ and $B \subseteq C$ we have $A \subseteq C$. The last two statements lead at once to the result of Exercise 2. Exercise 3 is a very slight modification of the last statement.

**Exercise 2.** Let $A$, $B$, $C$ be classes such that $A \subseteq B$, $B \subseteq C$, $C \subseteq A$. Prove that $A = B = C$.

**Exercise 3.** Let $A$, $B$, $C$ be classes such that $A \subset B$, $B \subset C$. Prove that $A \subset C$.

There are many ways of establishing the equivalences in Exercise 4; one is to show in turn that (1) is equivalent to each of the other four statements.

**Exercise 4.** Let $A$ and $B$ be classes. Prove that the following five statements are equivalent: (1) $A \subseteq B$; (2) $A \cup B = B$; (3) $A \cap B = A$; (4) $(\sim B) \subseteq (\sim A)$; (5) $A \cap (\sim B) = \emptyset$.

To establish the results of Exercises 5 and 6 we use the standard technique for proving that one class $X$ is included in another class $Y$: namely, we begin by saying "Let $x$ be any element of $X$" and try, using the defining properties of $X$ and $Y$, to show that $x \in Y$.

**Exercise 5.** Let $A$ and $B$ be classes such that $A \subseteq B$. Prove that $\bigcup A \subseteq \bigcup B$ and $\bigcap B \subseteq \bigcap A$.

**Exercise 6.** Let $a$ be a set, $B$ a class such that $a \in B$. Prove that $a \subseteq \bigcup B$ and $\bigcap B \subseteq a$.

For every class $A$ we define the **power class** $\mathbf{P}(A)$ of $A$ to be the class of all subsets of $A$, i.e. $\mathbf{P}(A) = \{x : x \subseteq A\}$.

**Axiom 3 (Power Set Axiom).** For every set $x$ there exists a set $y$ such that $u \in y$ if and only if $u \subseteq x$.

The Power Set Axiom thus asserts that every subclass $u$ of a set $x$ is actually a set (since it is an element of the set $y$ whose existence is asserted by the axiom) and furthermore that the power class $\mathbf{P}(x)$ of a set $x$ is also a set (and so is usually referred to as the **power set** of $x$).

The second and third assertions of Exercise 7 are established by the standard procedure for proving two classes equal: we show that each class is included in the other; the first assertion is best tackled by contradiction. To prove the result of Exercise 8 let $a$ be any element of $A$ and use Exercise 6.

**Exercise 7.** Prove that $\bigcap V = \emptyset$, $\bigcup V = V$ and $\mathbf{P}(V) = V$.

**Exercise 8.** Prove that if $A$ is a non-null class then $\bigcap A$ is a set.

Let $R = \{x : x \notin x\}$, the **Russell class**. This is a proper class, since if $R$ were a set we should have $R \in R$ if and only if $R \notin R$, which is a contradiction. It follows that $V$ is a proper class, since if $V$ were a set then $R$ (being a subclass of $V$) would also be a set.

**Axiom 4 (Pairing Axiom).** For all sets $x$ and $y$ the class $\{z : (z = x) \vee (z = y)\}$ is a set.

The set $\{z : (z = x) \vee (z = y)\}$ is denoted by $\{x, y\}$ and such a set is called an **unordered pair**. If $x = y$ the unordered pair $\{x, y\}$ is denoted by $\{x\}$ and is called **singleton** $x$.

**Axiom 5 (Union Axiom).** For every set $x$ the class $\bigcup x$ is a set.

Exercise 9 is a straightforward consequence of the definitions of the terms involved.

**Exercise 9.** Prove that for all sets $x$ and $y$ we have $x \cup y = \bigcup \{x, y\}$ and $\{x\} \cup \{y\} = \{x, y\}$.

# Chapter 2

# RELATIONS

Let $a$ and $b$ be sets. Then the set $\{\{a\},\{a,b\}\}$ is denoted by $(a,b)$ and is called the **ordered pair** with **first coordinate** $a$ and **second coordinate** $b$. Let $A$ and $B$ be classes; then the **Cartesian product** of $A$ and $B$ is the class

$$A \times B = \{t : (\exists\, x)(\exists\, y)((x \in A) \wedge (y \in B) \wedge (t = (x,y)))\},$$

i.e. $A \times B$ is the class of all ordered pairs with first coordinate in $A$ and second coordinate in $B$.

If $P(x,y)$ is an open sentence involving the free variables $x$ and $y$ we shall allow ourselves to write

$$\{(x,y) : P(x,y)\}$$

as an abbreviation for

$$\{t : (\exists\, x)(\exists\, y)((t = (x,y)) \wedge P(x,y))\}.$$

So we can abbreviate the definition of $A \times B$ to

$$A \times B = \{(x,y) : (x \in A) \wedge (y \in B)\}.$$

Suppose $A$ and $B$ are sets. For each element $a$ of $A$ and each element $b$ of $B$ we have $\{a\} \in \mathbf{P}(A \cup B)$ and $\{a,b\} \in \mathbf{P}(A \cup B)$. So $\{\{a\},\{a,b\}\} \subseteq \mathbf{P}(A \cup B)$ and hence $(a,b) \in \mathbf{P}(\mathbf{P}(A \cup B))$. Thus $A \times B \subseteq \mathbf{P}(\mathbf{P}(A \cup B))$. So $A \times B$ is a set.

**Exercise 10.** Let $a$, $b$, $c$, $d$ be sets. Prove that $(a,b) = (c,d)$ if and only if $a = c$ and $b = d$.

The *if* part is clear.

For the *only if* part distinguish the cases (1) where $a = b$ and (2) where $a \neq b$ and so $\{a, b\} \neq \{a\}$.

Exercises 11 and 12 are straightforward.

**Exercise 11.** Let $A$ and $B$ be classes. Prove that $A \times B \neq \emptyset$ if and only if $A \neq \emptyset$ and $B \neq \emptyset$.

**Exercise 12.** Let $A$ and $B$ be non-empty classes; let $A'$ and $B'$ be classes. Prove that $A \times B \subseteq A' \times B'$ if and only if $A \subseteq A'$ and $B \subseteq B'$.

**Exercise 13.** Let $A$, $B$, $C$ be classes. Prove

(1) $A \times (B \cup C) = (A \times B) \cup (A \times C)$;

(2) $A \times (B \cap C) = (A \times B) \cap (A \times C)$;

(3) $A \times (B \sim C) = (A \times B) \sim (A \times C)$.

All three parts of Exercise 13 can be established by a chain of equivalences starting with

$$(a, x) \in \text{left hand side} \iff$$

and ending with

$$\iff (a, x) \in \text{right hand side}.$$

Exercise 14 follows easily from the definition of the ordered pair $(a, b)$.

**Exercise 14.** Let $a$ and $b$ be sets. Prove

(1) $\bigcap \bigcap (a, b) = a$ and

(2) $\left(\bigcap \bigcup (a, b)\right) \cup \left(\bigcup \bigcup (a, b) \sim \bigcup \bigcap (a, b)\right) = b$.

A **relation** is a class of ordered pairs.

Let $R$ be a relation. We define the **domain** and **range** of $R$ to be the classes Dom $R$ and Range $R$ given by

$$\begin{aligned} \text{Dom } R &= \{x : (\exists y)((x, y) \in R)\}, \\ \text{Range } R &= \{y : (\exists x)((x, y) \in R)\}. \end{aligned}$$

If $R$ is a relation and $(x, y) \in R$ we say that $x$ is $R$-**related** to $y$ and that $y$ is an $R$-**relative** of $x$. Thus Dom $R$ is the class of all sets which *have* $R$-relatives and Range $R$ is the class of all sets which *are* $R$-relatives.

**Exercise 15.** Prove that if a relation $R$ is a set then Dom $R$ and Range $R$ are also sets.

Show that Dom $R$ and Range $R$ are included in $\bigcup\bigcup R$ and use the Union Axiom and the Power Set Axiom.

If $A$ and $B$ are classes then a **relation between** $A$ **and** $B$ is a subclass of $A \times B$, i.e. a relation $R$ such that Dom $R \subseteq A$ and Range $R \subseteq B$. A **relation on** a class $A$ is a subclass of $A \times A$.

If $R$ is a relation, the **inverse** of $R$ is the relation $R^{-1}$ given by

$$R^{-1} = \{(x, y) : (y, x) \in R\}.$$

If $R$ is a relation between $A$ and $B$ then $R^{-1}$ is a relation between $B$ and $A$; clearly Dom $R^{-1}$ = Range $R$ and Range $R^{-1}$ = Dom $R$.

If $R$ and $S$ are relations, the **composition** of $R$ and $S$ is the relation $S \circ R$ given by

$$S \circ R = \{(x, z) : (\exists y)(((x, y) \in R) \wedge ((y, z) \in S))\}.$$

If $R$ is a relation between $A$ and $B$ and $S$ is a relation between $B$ and $C$ then clearly $S \circ R$ is a relation between $A$ and $C$ and we have Dom $(S \circ R) \subseteq$ Dom $R$ and Range $(S \circ R) \subseteq$ Range $S$.

Exercises 16, 17 and 18 follow by completing the chains of equivalences

$$\begin{aligned}
(z, x) \in (S \circ R)^{-1} &\iff \cdots \iff (z, x) \in R^{-1} \circ S^{-1}, \\
(x, t) \in T \circ (S \circ R) &\iff \cdots \iff (x, t) \in (T \circ S) \circ R, \\
(x, y) \in (R \cap S)^{-1} &\iff \cdots \iff (x, y) \in R^{-1} \cap S^{-1}, \\
(x, y) \in (R \cup S)^{-1} &\iff \cdots \iff (x, y) \in R^{-1} \cup S^{-1}.
\end{aligned}$$

**Exercise 16.** Let $R$ and $S$ be relations. Prove that $(S \circ R)^{-1} = R^{-1} \circ S^{-1}$.

**Exercise 17.** Let $R$, $S$, $T$ be relations. Prove that $T \circ (S \circ R) = (T \circ S) \circ R$.

In view of this result we may omit the parentheses and write both sides of the last equation as $T \circ S \circ R$.

**Exercise 18.** Let $R$ and $S$ be relations. Prove that we have $(R \cap S)^{-1} = R^{-1} \cap S^{-1}$ and $(R \cup S)^{-1} = R^{-1} \cup S^{-1}$.

Let $R$ be a relation, $A$ any class. Then the **image** of $A$ **under** $R$ is the class consisting of all $R$-relatives of all members of $A$. We denote this class by $R^{\rightarrow}(A)$. Thus

$$R^{\rightarrow}(A) = \{y : (\exists\, x)((x \in A) \wedge ((x, y) \in R))\}.$$

Again let $R$ be a relation and let $B$ be any class. Then the **inverse image** of $B$ **under** $R$ is the class $(R^{-1})^{\rightarrow}(B)$, which we write $R^{\leftarrow}(B)$. Thus

$$R^{\leftarrow}(B) = \{x : (\exists\, y)((y \in B) \wedge ((x, y) \in R))\}.$$

Most authors use the notations $R(A)$ or $R[A]$ for the image $R^{\rightarrow}(A)$ of $A$ under $R$ and $R^{-1}(B)$ or $R^{-1}[B]$ for the inverse image $R^{\leftarrow}(B)$ of $B$ under $R$.

Exercises 19 and 20, being *if and only if* statements, each fall into two parts.

In Exercise 19 we suppose first that $A \subseteq \operatorname{Dom} R$, take any element $a$ of $A$ and prove that $a \in R^{\leftarrow}(R^{\rightarrow}(A))$; secondly we suppose that $A \subseteq R^{\leftarrow}(R^{\rightarrow}(A))$, take any element $a$ of $A$ and show that $a \in \operatorname{Dom} R$.

Similarly in Exercise 20 we suppose first that $\operatorname{Dom} S \subseteq \operatorname{Dom} R$, take any ordered pair $(a, b)$ in $S$ and prove that $(a, b) \in S \circ R^{-1} \circ R$; then we suppose that $S \subseteq S \circ R^{-1} \circ R$, take any element $a$ of $\operatorname{Dom} S$ and prove that $a \in \operatorname{Dom} R$, in each case using the definitions of domain and composition of relations.

**Exercise 19.** Let $R$ be a relation, $A$ a class. Prove that $A \subseteq \operatorname{Dom} R$ if and only if $A \subseteq R^{\leftarrow}(R^{\rightarrow}(A))$.

**Exercise 20.** Let $R$ and $S$ be relations. Prove that we have $\operatorname{Dom} S \subseteq \operatorname{Dom} R$ if and only if $S \subseteq S \circ R^{-1} \circ R$.

Exercise 21 and the first part of Exercise 22 are established by setting up chains of equivalences

$$\begin{aligned}(a,c) \in R \circ (A \times B) &\iff \cdots \iff (a,c) \in A \times R^{\rightarrow}(B),\\ (b,d) \in (C \times D) \circ R &\iff \cdots \iff (b,d) \in R^{\leftarrow}(C) \times D,\\ (x,y) \in (R')^{-1} &\iff \cdots \iff (x,y) \in (R^{-1})'.\end{aligned}$$

For the last two parts of Exercise 22 we have to construct chains of implications

$$\begin{aligned}(x,y) \in R \circ (R^{-1})' &\implies \cdots \implies (x,y) \in (D_B)',\\ (x,y) \in (R^{-1})' \circ R &\implies \cdots \implies (x,y) \in (D_A)'.\end{aligned}$$

**Exercise 21.** Let $A$, $B$, $C$, $D$ be classes, $R$ a relation between $B$ and $C$. Prove that

(1) $R \circ (A \times B) = A \times R^{\rightarrow}(B)$;

(2) $(C \times D) \circ R = R^{\leftarrow}(C) \times D$.

**Exercise 22.** For each relation $S$ we define the class $S' = (\text{Dom } S \times \text{ Range } S) \sim S$. Prove that for every relation $R$ we have $(R')^{-1} = (R^{-1})'$ and that for all classes $A$, $B$ such that $A \supseteq \text{Dom } R$ and $B \supseteq \text{Range } R$ we have $R \circ (R^{-1})' \subseteq (D_B)'$ and $(R^{-1})' \circ R \subseteq (D_A)'$ where we define

$$\begin{aligned}D_A &= \{t : (\exists a)((a \in A) \wedge (t = (a,a)))\},\\ D_B &= \{t : (\exists b)((b \in B) \wedge (t = (b,b)))\}.\end{aligned}$$

or, informally, $D_A = \{(a,a) : a \in A\}$ and $D_B = \{(b,b) : b \in B\}$.

# Chapter 3

# FUNCTIONAL RELATIONS AND MAPPINGS

A relation $R$ is said to be **functional** if each element of its domain has exactly one $R$-relative; a functional relation is also called a **function**. If $R$ is a functional relation then for each element $a$ of its domain we denote the unique $R$-relative of $a$ by $R(a)$. Clearly if $R$ and $S$ are functional relations then so is $S \circ R$.

**Exercise 23.** Prove that a relation $R$ is functional if and only if $R^{\rightarrow}(R^{\leftarrow}(Y)) \subseteq Y$ for every class $Y$.

Again we are dealing with a two-part *if and only if* result. So we suppose first that $R$ is functional, take any class $Y$ and any element $b$ of $R^{\rightarrow}(R^{\leftarrow}(Y))$ and then try to prove, using the definitions of $R^{\rightarrow}$, $R^{\leftarrow}$ and functionality, that $b \in Y$. For the converse we suppose that for every class $Y$ we have $R^{\rightarrow}(R^{\leftarrow}(Y)) \subseteq Y$. We have to prove, under this hypothesis, that if $a$ is any element of the domain of $R$ and $(a, b_1)$, $(a, b_2)$ are in $R$ then $b_1 = b_2$. So we must choose an appropriate particular class $Y$ to which to apply the hypothesis.

**Exercise 24.** Let $R$ be a relation. Prove that the following three statements are equivalent:

(1) $R$ is functional;

(2) for all classes $X$ and $Y$ we have $R^{\leftarrow}(X \cap Y) = R^{\leftarrow}(X) \cap R^{\leftarrow}(Y)$;

(3) for all classes $X$ and $Y$ such that $X \cap Y = \emptyset$ we have $R^{\leftarrow}(X) \cap R^{\leftarrow}(Y) = \emptyset$.

To establish the equivalences asserted in Exercise 24 we prove the three implications (a) (1) $\Longrightarrow$ (2), (b) (2) $\Longrightarrow$ (3), (c) (3) $\Longrightarrow$ (1) or the equivalent contrapositive form $\neg(1) \Longrightarrow \neg(3)$.

For (a) we suppose that $R$ is functional; take any two classes $X$ and $Y$ and try to prove first that $R^{\leftarrow}(X \cap Y) \subseteq R^{\leftarrow}(X) \cap R^{\leftarrow}(Y)$, which actually holds for every relation $R$, and then that $R^{\leftarrow}(X) \cap R^{\leftarrow}(Y) \subseteq R^{\leftarrow}(X \cap Y)$, which requires the functional property of $R$.

In (b) we see that (3) is just a special case of (2).

In (c) we assume that (1) does not hold; thus there is an element $a$ of the domain of $R$ which has two distinct $R$-relatives, say $b_1$ and $b_2$. We use these to form two classes $X$ and $Y$ with $X \cap Y = \emptyset$ for which $R^{\leftarrow}(X) \cap R^{\leftarrow}(Y)$ is non-empty; thus (3) does not hold.

**Axiom 6 (Replacement Axiom).** For every functional relation $R$, if the domain of $R$ is a set then the range of $R$ is also a set.

A **mapping** is an ordered pair $((A, B), R)$ where $A$ and $B$ are sets and $R$ is a functional relation between $A$ and $B$ such that $\operatorname{Dom} R = A$. If $f = ((A, B), R)$ is a mapping we say that $f$ is a **mapping from** $A$ **to** $B$; we call $A$ the **domain** of $f$, $B$ the **codomain** of $f$ and $R$ the **graph** of $f$. If $f$ is a mapping with domain $A$ and codomain $B$ we often write $f : A \to B$. If $a$ is any element of the set $A$ then the set $R^{\rightarrow}(\{a\})$ consists of a single element of $B$ which we denote by $f(a)$; we call it the **image of** $a$ **under** $f$ or the **value of** $f$ **at** $a$.

It is clear from the definition of the term "mapping" that in order to describe a mapping $f$ we must give the domain $A$ and codomain $B$ of $f$ and also, for each element $a$ of the domain we must describe the unique element $b_a$ of the codomain such that $(a, b_a)$ belongs to the graph of $f$, i.e. we must describe for each element $a$ of $A$ its image under $f$ in $B$.

Let $f = ((A, B), R)$ be a mapping from $A$ to $B$. For each subset $X$ of $A$ we denote the subset $R^{\rightarrow}(X)$ of $B$ by $f^{\rightarrow}(X)$; in particular, if $a$ is any element of $A$ we have $f^{\rightarrow}(\{a\}) = \{f(a)\}$. For each subset $Y$ of $B$ we denote the subset $R^{\leftarrow}(Y)$ of $A$ by $f^{\leftarrow}(Y)$.

Let $f = ((A, B), R)$ be a mapping from $A$ to $B$; let $A_1$ be a subset of $A$. Then the **restriction** of $f$ to $A_1$ is the mapping $f \mid A_1 = ((A_1, B), R \cap (A_1 \times B))$.

**Exercise 25.** Let $f = ((A, B), R)$ and $f' = ((A', B'), R')$ be mappings. Prove that $f = f'$ if and only if $A = A'$, $B = B'$ and

$f(a) = f'(a)$ for all $a$ in $A$.

**Exercise 26.** Let $A$ and $B$ be sets. Prove that the class $\mathrm{Map}(A, B) = \{f : f$ is a mapping from $A$ to $B\}$ is a set.

Exercise 26 follows when we notice that we have $\mathrm{Map}(A, B) \subseteq \{(A, B)\} \times \mathbf{P}(A \times B)$ and use the Pairing Axiom, the fact that the Cartesian product of two sets is a set and the Power Set Axiom, each twice.

**Exercise 27.** Prove that if $A$ and $B$ are sets then $\mathrm{Map}(A, B) = \emptyset$ if and only if $A$ is non-empty and $B$ is empty.

Consider the cases (a) $A \neq \emptyset$ and $B = \emptyset$; (b) $A = \emptyset$; (c) $A \neq \emptyset$ and $B \neq \emptyset$ and look in each case for functional relations between $A$ and $B$.

Let $f = ((A, B), R)$ and $g = ((B, C), S)$ be mappings. Then clearly $((A, C), S \circ R)$ is also a mapping, which we denote by $g \circ f$ and call the **composition** or **composed mapping** of $f$ and $g$. For each element $a$ of $A$ we have $(g \circ f)(a) = g(f(a))$.

**Exercise 28.** Let $A$, $B$, $C$, $D$ be sets; let $f$, $g$, $h$ be mappings from $A$ to $B$, $B$ to $C$, $C$ to $D$ respectively. Prove that $h \circ (g \circ f) = (h \circ g) \circ f$.

This follows immediately from Exercise 17.

In view of this result we may drop the parentheses and write both mappings $h \circ (g \circ f)$ and $(h \circ g) \circ f$ simply as $h \circ g \circ f$.

Let $A$ be any set; let $D_A = \{x : (\exists a)((a \in A) \wedge (x = (a, a)))\}$ (we call $D_A$ the **diagonal** of $A \times A$). Then $D_A$ is a functional relation between $A$ and $A$ with domain $A$. The mapping $I_A = ((A, A), D_A)$ from $A$ to $A$ is called the **identity mapping** of $A$. Clearly we have $I_A(a) = a$ for each element $a$ of $A$.

**Exercise 29.** Prove that for every mapping $f$ with domain $A$ we have $f \circ I_A = f$ and for every mapping $g$ with codomain $A$ we have $I_A \circ g = g$.

This follows immediately from the definition of $I_A$.

Let $A$ and $B$ be sets; let

$$\begin{aligned} P_1 &= \{z : (\exists a)(\exists b)((a \in A) \wedge (b \in B) \wedge (z = ((a,b),a)))\}, \\ P_2 &= \{z : (\exists a)(\exists b)((a \in A) \wedge (b \in B) \wedge (z = ((a,b),b)))\}. \end{aligned}$$

Then $P_1$ and $P_2$ are functional relations with domain $A \times B$. The mappings $\pi_1 = ((A \times B, A), P_1)$ and $\pi_2 = ((A \times B, B), P_2)$ from $A \times B$ to $A$ and $B$ respectively are called the **first** and **second projections** from $A \times B$. For each ordered pair $(a, b)$ in $A \times B$ we have $\pi_1((a,b)) = a$ and $\pi_2((a,b)) = b$. We usually write $\pi_i(a,b)$ instead of $\pi_i((a,b))$ ($i = 1$, 2).

Let $f = ((A, B), R)$ be a mapping.

Then $f$ is said to be **surjective** or to be a **surjection** if we have Range $R = B$, i.e. if every element of $B$ is the image under $f$ of at least one element of $A$.

Next, $f$ is said to be **injective** or to be an **injection** if the inverse relation $R^{-1}$ is functional. Thus $f$ is an injection if and only if for every element $b$ in Range $R$ there is exactly one element $a$ of $A$ such that $(b, a) \in R^{-1}$, i.e. $(a, b) \in R$ and so $b = f(a)$. So $f$ is injective if and only if each element of the range of $R$ is the image under $f$ of exactly one element of $A$. Again, $f$ is injective if and only if whenever $f(a_1) = f(a_2)$, where $a_1, a_2 \in A$, then $a_1 = a_2$.

The mapping $f$ is said to be **bijective** or to be a **bijection** if it is both injective and surjective. If $A$ and $B$ are sets we say that $A$ **is equipotent to** $B$ or that $A$ **is equinumerous with** $B$ if there exists a bijective mapping from $A$ to $B$.

Let $f = ((A, B), R)$ be a mapping. Then $((B, A), R^{-1})$ is a mapping if and only if $f$ is bijective; in this case we write $((B, A), R^{-1}) = f^{-1}$ and call $f^{-1}$ the **inverse mapping** of $f$. Clearly $f^{-1}$ is a bijection from $B$ to $A$ with inverse $f$.

**Exercise 30.** Let $f$ be a mapping from a non-empty set $A$ to a non-empty set $B$. Prove that the following statements are equivalent:

(1) $f$ is injective;

(2) there exists a mapping $g$ from $B$ to $A$ such that $g \circ f = I_A$;

(3) for every set $X$ and all mappings $h_1$, $h_2$ from $X$ to $A$ such that $f \circ h_1 = f \circ h_2$, we have $h_1 = h_2$.

Here we proceed by proving the implications (a) (1) $\implies$ (2), (b) (2) $\implies$ (3), (c) $\neg(1) \implies \neg(3)$.

For (a) we suppose that $f$ is injective and have to construct a mapping $g$ from $B$ to $A$, in particular we have to define $g(b)$ for every element $b$ of $B$. Distinguish the cases where $b$ is and is not in the image of $A$ under $f$.

In (b) we simply compose the mapping $g$ of (1) with both sides of the equation $f \circ h_1 = f \circ h_2$.

In (c), if $f$ is not injective there are elements $a_1$, $a_2$ of $A$ such that $a_1 \neq a_2$ but $f(a_1) = f(a_2)$. Use $a_1$ and $a_2$ to construct mappings $h_1$ and $h_2$ from an appropriate set such that $h_1 \neq h_2$ but $f \circ h_1 = f \circ h_2$.

We sometimes express the properties (2) and (3) of injective mappings established in Exercise 30 by saying that injective mappings are **left invertible** and **left cancellable.** Once we have introduced the Axiom of Choice in Chapter 4 we shall be able to prove an analogous result to Exercise 30 for surjective mappings and right invertibility and right cancellability (see Exercise 47).

**Exercise 31.** Let $A$, $B$, $C$ be non-empty sets; let $f$ be a mapping from $A$ to $B$, $g$ a mapping from $B$ to $C$; let $h = g \circ f$. Prove

(1) if $f$ and $g$ are injective, then $h$ is injective;
(2) if $f$ and $g$ are surjective, then $h$ is surjective;
(3) if $h$ is injective, then $f$ is injective;
(4) if $h$ is surjective, then $g$ is surjective;
(5) if $h$ is injective and $f$ is surjective, then $g$ is injective;
(6) if $h$ is surjective and $g$ is injective, then $f$ is surjective.

For (1) and (3) we can use either the definition of injectivity or the equivalent condition (3) of Exercise 29. (2) and (4) follow easily from the definition of surjectivity. In (5) we suppose we have elements $b_1$, $b_2$ of $B$ such that $g(b_1) = g(b_2)$ and try to deduce that $b_1 = b_2$. In (6) we take any element $b$ of $B$, consider $g(b)$ in $C$ and use the surjectivity of the mapping $h$.

**Exercise 32.** Let $A$, $B$, $C$ be non-empty sets; let $f$, $g$, $h$ be mappings from $A$ to $B$, $B$ to $C$, $C$ to $A$ respectively. If $h \circ g \circ f$ and $g \circ f \circ h$ are injections and $f \circ h \circ g$ is a surjection, prove that $f$, $g$ and $h$ are all bijections.

The results of Exercise 32 follow by using appropriate parts of Exercise 31.

**Exercise 33.** Let $f$ be a mapping from a set $A$ to a set $B$. Prove
(1) $f^{\rightarrow}(\emptyset) = \emptyset$;
(2) for every element $x$ of $A$ we have $f^{\rightarrow}(\{x\}) = \{f(x)\}$;
(3) for all subsets $X_1$, $X_2$ of $A$ such that $X_1 \subseteq X_2$ we have

$$f^{\rightarrow}(X_1) \subseteq f^{\rightarrow}(X_2);$$

(4) for all subsets $X_1$, $X_2$ of $A$ we have

$$f^{\rightarrow}(X_1 \cup X_2) = f^{\rightarrow}(X_1) \cup f^{\rightarrow}(X_2);$$

(5) for all subsets $X_1$, $X_2$ of $A$ we have

$$f^{\rightarrow}(X_1 \cap X_2) \subseteq f^{\rightarrow}(X_1) \cap f^{\rightarrow}(X_2);$$

(6) if $f$ is injective then for all subsets $X_1$, $X_2$ of $A$ we have

$$f^{\rightarrow}(X_1 \cap X_2) = f^{\rightarrow}(X_1) \cap f^{\rightarrow}(X_2);$$

(7) for all subsets $Y$ of $B$ we have $f^{\leftarrow}(B \sim Y) = A \sim f^{\leftarrow}(Y)$;
(8) for all subsets $Y_1$, $Y_2$ of $B$ such that $Y_1 \subseteq Y_2$ we have

$$f^{\leftarrow}(Y_1) \subseteq f^{\leftarrow}(Y_2);$$

(9) for all subsets $Y_1$, $Y_2$ of $B$ we have

$$f^{\leftarrow}(Y_1 \cup Y_2) = f^{\leftarrow}(Y_1) \cup f^{\leftarrow}(Y_2);$$

(10) for all subsets $Y_1$, $Y_2$ of $B$ we have

$$f^{\leftarrow}(Y_1 \cap Y_2) = f^{\leftarrow}(Y_1) \cap f^{\leftarrow}(Y_2);$$

(11) if $f$ is bijective then for all subsets $X$ of $A$ we have

$$f^{\rightarrow}(A \sim X) = B \sim f^{\rightarrow}(X);$$

(12) for every subset $X$ of $A$ we have $f^{\leftarrow}(f^{\rightarrow}(X)) \supseteq X$;
(13) $f$ is injective if and only if $f^{\leftarrow}(f^{\rightarrow}(X)) = X$ for all subsets $X$ of $A$;
(14) for every subset $Y$ of $B$ we have $f^{\rightarrow}(f^{\leftarrow}(Y)) \subseteq Y$;

(15) $f$ is surjective if and only if $f^{\rightarrow}(f^{\leftarrow}(Y)) = Y$ for all subsets $Y$ of $B$.

Parts (1), (2) and (3) of Exercise 33 are straightforward.

In part (4) the inclusion $f^{\rightarrow}(X_1) \cup f^{\rightarrow}(X_2) \subseteq f^{\rightarrow}(X_1 \cup X_2)$ follows from part (3); the reverse inclusion is obtained by considering any element $y$ of $f^{\rightarrow}(X_1 \cup X_2)$.

Part (5) follows at once from part (3).

To establish part (6) we see that, in view of part (5), we have only to prove that if $f$ is injective then $f^{\rightarrow}(X_1) \cap f^{\rightarrow}(X_2) \subseteq f^{\rightarrow}(X_1 \cap X_2)$; so let $y$ be any element in the left hand side and use the injectivity to show that $y$ is in the right hand side.

Parts (7) and (8) are immediate.

In parts (9) and (10) the inclusions $f^{\leftarrow}(Y_1) \cup f^{\leftarrow}(Y_2) \subseteq f^{\leftarrow}(Y_1 \cup Y_2)$ and $f^{\leftarrow}(Y_1 \cap Y_2) \subseteq f^{\leftarrow}(Y_1) \cap f^{\leftarrow}(Y_2)$ follow from part (8); the reverse inclusions are obtained by taking any element $x$ of $f^{\leftarrow}(Y_1 \cup Y_2)$ or $f^{\leftarrow}(Y_1) \cap f^{\leftarrow}(Y_2)$ respectively and considering $f(x)$.

For part (11) take an element $y$ in $f^{\rightarrow}(A \sim X)$ and use the injectivity of $f$ to show that $y$ cannot be in $f^{\rightarrow}(X)$; then take an element $y$ of $B \sim f^{\rightarrow}(X)$ and use the surjectivity of $f$ to show that $y$ is in $f^{\rightarrow}(A \sim X)$.

Parts (12) and (14) follow at once from the definitions of $f^{\rightarrow}$ and $f^{\leftarrow}$.

For part (13), if $f$ is injective, take any subset $X$ of $A$ and any element $t$ of $f^{\leftarrow}(f^{\rightarrow}(X))$; by considering $f(t)$ show that $t$ must lie in $X$. Conversely, if $f^{\leftarrow}(f^{\rightarrow}(X)) = X$ for all subsets $X$ of $A$ prove $f$ is injective by supposing we have $f(a_1) = f(a_2)$ and applying the condition to a suitably constructed subset $X$ of $A$.

Finally, for part (15), if $f$ is surjective, let $Y$ be any subset of $B$, $y$ any element of $Y$ and use the surjectivity of $f$ to show that $y \in f^{\rightarrow}(f^{\leftarrow}(Y))$. Conversely, if $f^{\rightarrow}(f^{\leftarrow}(Y)) = Y$ for all subsets $Y$ of $B$, prove that each element $b$ of $B$ is an image under $f$ by applying the condition to an appropriate subset $Y$ of $B$.

**Exercise 34.** Let $A$, $B$, $C$ be non-empty sets, $f$ a mapping from $A$ to $C$, $g$ a mapping from $B$ to $C$. If $g$ is injective, prove that there exists a mapping $h$ from $A$ to $B$ such that $f = g \circ h$ if and only if $f^{\rightarrow}(A) \subseteq g^{\rightarrow}(B)$. Show that if this condition is satisfied then the mapping $h$ is unique.

It is clear that if $f = g \circ h$ then $f^{\rightarrow}(A) \subseteq g^{\rightarrow}(B)$ whether $g$ is injective or not. If $g$ is injective the converse (and the uniqueness) follow using Exercise 30.

**Exercise 35.** Let $A_1$ and $A_2$ be sets, $\pi_1$ and $\pi_2$ the first and second projections from $A_1 \times A_2$ onto $A_1$ and $A_2$ respectively. Prove that for every set $X$ and all mappings $f_1$, $f_2$ from $X$ to $A_1$, $A_2$ respectively, there is a unique mapping $f$ from $X$ to $A_1 \times A_2$ such that $\pi_1 \circ f = f_1$ and $\pi_2 \circ f = f_2$.

Clearly the mapping $f$ from $X$ to $A_1 \times A_2$ defined by setting $f(x) = (f_1(x), f_2(x))$ for all elements $x$ of $X$ satisfies the required condition; the uniqueness is almost immediate.

**Exercise 36.** Let $A_1$, $A_2$, $B_1$, $B_2$ be sets; let $f_1$ be a mapping from $A_1$ to $B_1$ and $f_2$ a mapping from $A_2$ to $B_2$. Define a "natural" mapping $f$ from $A_1 \times A_2$ to $B_1 \times B_2$ and prove that if $f_1$ and $f_2$ are bijections then so is $f$.

In Exercise 36 we have to define $f((a_1, a_2))$ for all elements $(a_1, a_2)$ of $A_1 \times A_2$. There is an obvious ("natural") way to do this.

**Exercise 37.** Let $A_1$, $A_2$, $B_1$, $B_2$ be sets; let $f$ be a mapping from $A_2$ to $A_1$ and $g$ a mapping from $B_1$ to $B_2$. Define a "natural" mapping from $\mathrm{Map}(A_1, B_1)$ to $\mathrm{Map}(A_2, B_2)$.

Here we have to define for every mapping $k$ from $A_1$ to $B_1$ a mapping $h(k)$ from $A_2$ to $B_2$. There is a natural way to do this using $k$ and the given mappings $f$ and $g$.

**Exercise 38.** Let $A$, $B$, $C$ be sets. Define a "natural" mapping $F$ from $\mathrm{Map}(A, \mathrm{Map}(B, C))$ to $\mathrm{Map}(B \times A, C)$ and prove that it is a bijection.

For each mapping $f$ from $A$ to $\mathrm{Map}(B, C)$ we have to define a mapping $F(f)$ from $B \times A$ to $C$; so we must decide for each element $(b, a)$ of $B \times A$ which element of $C$ is to serve as $(F(f))(b, a)$. Since $f$ can be applied to $a$ to give a mapping $f(a)$ from $B$ to $C$ and this mapping can be applied to $b$ to give an element of $C$, it is natural to set $(F(f))(b, a) = (f(a))(b)$ for each ordered pair $(b, a)$.

The proof that $F$ is bijective is routine.

**Exercise 39.** Let $A$, $B$, $C$ be sets. Define a "natural" mapping from $\mathrm{Map}(C, A \times B)$ to $\mathrm{Map}(C, A) \times \mathrm{Map}(C, B)$ and prove it is a bijection.

For each mapping $f$ from $C$ to $A \times B$ we have to define an ordered pair whose first coordinate is a mapping from $C$ to $A$ and whose second coordinate is a mapping from $C$ to $B$. There is an obvious way to do this.

# Chapter 4

# FAMILIES OF SETS

Let $I$ and $A$ be classes, $F$ a functional relation between $I$ and $A$. Then $F$ is sometimes called a **family of elements of $A$ indexed by $I$** (or **with $I$ as index class**) and we write $(F(i))_{i\in I}$ instead of $F$. In particular, if $E$ is a set then a family of elements of $\mathbf{P}(E)$ is called a **family of subsets** of $E$ indexed by $I$. If $F$ is such a family and we write $X_i = F(i)$ for each element $i$ in $I$ then we denote the family $F$ by $(X_i)_{i\in I}$.

If $X$ is a set of subsets of a set $E$ then the diagonal $D_X$ is a family of subsets of $E$ which may sometimes be denoted by $(x)_{x\in X}$.

Let $F = (X_i)_{i\in I}$ be a family of subsets of a set $E$. We define the **union** of the family to be

$$\bigcup_{i\in I} X_i = \{x : (\exists i)((i \in I) \wedge (x \in X_i))\}.$$

Thus $x$ belongs to the union of the family $(X_i)_{i\in I}$ if and only if it belongs to at least one of the sets $X_i$ with $i$ in $I$. Of course if $X$ is a set of subsets of $E$ the union of the corresponding family $D_X$ coincides with the union of the set $X$ as previously defined: $\bigcup_{x\in X} x = \bigcup X$.

Again let $(X_i)_{i\in I}$ be a family of subsets of a set $E$. The **intersection** of this family is defined to be

$$\bigcap_{i\in I} X_i = \{x : (x \in E) \wedge (\forall i)((i \in I) \Longrightarrow (x \in X_i))\}.$$

So $x$ belongs to the intersection of the family $(X_i)_{i\in I}$ if and only if it belongs to all the sets $X_i$ with $i$ in $I$.

We notice that if $I$ is empty then we have $\bigcap_{i\in I} X_i = E$. If $X$ is a non-empty set of subsets of $E$ the intersection of the corresponding family $D_X$ coincides with the intersection of the set $X$ as previously defined: $\bigcap_{x\in X} x = \bigcap X$.

Exercises 40, 41 and 42, which are all concerned with set equalities, are established in the standard way by showing that each of the two sets which are asserted to be equal is included in the other.

**Exercise 40.** Let $(X_i)_{i\in I}$ be a family of subsets of a set $E$. Let $f$ be a surjection from a set $K$ onto $I$. Prove that

$$\bigcup_{k\in K} X_{f(k)} = \bigcup_{i\in I} X_i$$

and

$$\bigcap_{k\in K} X_{f(k)} = \bigcap_{i\in I} X_i.$$

**Exercise 41.** Let $(X_i)_{i\in I}$ be a family of subsets of a set $E$. Suppose $I = \bigcup_{k\in K} J_k$. Prove that

$$\bigcup_{i\in I} X_i = \bigcup_{k\in K} \left(\bigcup_{i\in J_k} X_i\right)$$

and

$$\bigcap_{i\in I} X_i = \bigcap_{k\in K} \left(\bigcap_{i\in J_k} X_i\right).$$

**Exercise 42.** Let $(R_i)_{i\in I}$ be a family of relations between the sets $A$ and $B$; let $S$ be a relation between $B$ and a set $C$. Prove that $S \circ \left(\bigcup_{i\in I} R_i\right) = \bigcup_{i\in I}(S \circ R_i)$.

**Exercise 43.** Let $(F_i)_{i\in I}$ be a family of functional relations such that for every pair of indices $i$, $j$ in $I$ we have $F_i(x) = F_j(x)$ for every element $x$ of $\operatorname{Dom} F_i \cap \operatorname{Dom} F_j$. Prove that there exists a functional relation $F$ with domain $D = \bigcup_{i\in I} \operatorname{Dom} F_i$ such that $F(x) = F_i(x)$ for every element $x$ of $\operatorname{Dom} F_i$.

Consider the class $F$ of all ordered pairs $(x, y)$ such that $x$ is in the domain of some $F_i$ in the given family and $(x, y) \in F_i$.

**Exercise 44.** Let $(A_i)_{i\in I}$ be a family of sets indexed by a set $I$; for each index $i$ in $I$ let $f_i$ be a mapping from $A_i$ to $B$. Suppose that for every pair of indices $i$, $j$ in $I$ we have $f_i \,|\, (A_i \cap A_j) = f_j \,|\, (A_i \cap A_j)$.

Prove that there is a mapping $f$ from $A = \bigcup_{i \in I} A_i$ to $B$ such that $f \mid A_i = f_i$ for every index $i$ in $I$.

Apply the result of Exercise 43.

Let $(X_i)_{i \in I}$ be a family of sets with index set $I$. Let $X = \bigcup_{i \in I} X_i$. Then the **product** of the family $(X_i)_{i \in I}$ is the set

$$\prod_{i \in I} X_i = \{f : (f \in \mathrm{Map}(I, X)) \wedge (\forall i)((i \in I) \Longrightarrow (f(i) \in X_i))\}.$$

If we write $f(i) = x_i$ for each index $i$ in $I$ it is sometimes helpful to denote the element $f$ of $\prod_{i \in I} X_i$ by $\prod_{i \in I} x_i$.

For each index $j$ in $I$ we define a mapping $\pi_j$ from $\prod_{i \in I} X_i$ to $X_j$ by setting

$$\pi_j(f) = f(j) \quad \text{for every } f \text{ in } \prod_{i \in I} X_i.$$

The mapping $\pi_j$ is called the $j$**-th projection mapping** from $\prod_{i \in I} X_i$.

**Exercise 45.** Let $(X_i)_{i \in I}$ be a family of sets with index set $I$, $P$ their product and $A$ a set. Let $(f_i)_{i \in I}$ be a family of mappings from $A$ to $(X_i)_{i \in I}$ (i.e. for each index $i$ in $I$, $f_i$ is a mapping from $A$ to $X_i$). Prove that there is a unique mapping $f$ from $A$ to $P$ such that $\pi_i \circ f = f_i$ for each index $i$ in $I$.

To define a mapping $f$ from $A$ to $P$ we must give for every element $a$ of $A$ the corresponding element $f(a)$ of $P$. Since each of these elements $f(a)$ is a mapping from $I$ to $\bigcup_{i \in I} X_i$ we must define $(f(a))(i)$—which is an element of $X_i$—for each index $i$ in $I$. When we recall that we want $f$ to satisfy the conditions $\pi_i \circ f = f_i$ there is an obvious way to do this using the mappings $f_i$.

**Exercise 46.** Let $K$ be a non-empty set, $(J_k)_{k \in K}$ a family of non-empty sets indexed by $K$ and $I = \prod_{k \in K} J_k$. Let $((X_{i,k})_{i \in J_k})_{k \in K}$ be a family of families of sets. Prove that

$$\bigcup_{k \in K} \left( \bigcap_{i \in J_k} X_{i,k} \right) = \bigcap_{f \in I} \left( \bigcup_{k \in K} X_{f(k),k} \right).$$

Examine the special case where $K = \{1, 2\}$, $J_1 = \{1\}$, and $J_2 = \{1, 2\}$.

To prove the inclusion $\bigcup_{k\in K}\left(\bigcap_{i\in J_k} X_{i,k}\right) \subseteq \bigcap_{f\in I}\left(\bigcup_{k\in K} X_{f(k),k}\right)$ we have to show that for every element $x$ of the set on the left and every mapping $f$ in $I$ we have $x \in \bigcup_{k\in K} X_{f(k),k}$.

To prove the reverse inclusion it is convenient to suppose that $x$ is not in the set on the left; then $x$ fails to belong to each $\bigcap_{i\in J_k} X_{i,k}$ (for all $k$ in $K$) and so (for each index $k$) fails to belong to some $X_{j_k,k}$. Use the appropriate indices $j_k$ to form an element $f_0$ of $I$ and show that $x \notin \bigcup_{k\in K} X_{f_0(k),k}$.

Let $(E_i)_{i\in I}$ be a family of non-empty sets. By a **choice function** for this family we mean a mapping $f$ from $I$ to $\bigcup_{i\in I} E_i$ such that for every index $i$ in $I$ we have $f(i) \in E_i$; thus a choice function is an element of the product $\prod_{i\in I} E_i$. If we have $E_i \cap E_j = \emptyset$ for every pair of distinct elements $i$, $j$ of $I$ (we say that the family $(E_i)_{i\in I}$ is *pairwise disjoint*) then the range of a choice function for $(E_i)_{i\in I}$ is a subset of $\bigcup_{i\in I} E_i$ which contains exactly one element from each set of the family. Such a set is called a **selection set** for the (pairwise disjoint) family $(E_i)_{i\in I}$.

**Axiom 7 (Axiom of Choice).** Let $(E_i)_{i\in I}$ be a family of non-empty sets indexed by a set $I$. Then the product $\prod_{i\in I} E_i$ is non-empty.

Thus the Axiom of Choice asserts that for every family of non-empty sets there exists a choice function and so for every pairwise disjoint family of non-empty sets there exists a selection set.

**Exercise 47.** Let $f$ be a mapping from a non-empty set $A$ to a non-empty set $B$. Prove that the following statements are equivalent:

(1) $f$ is surjective;

(2) there exists a mapping $g$ from $B$ to $A$ such that $f \circ g = I_A$;

(3) for every set $Y$ and all mappings $h_1$, $h_2$ from $B$ to $Y$ such that $h_1 \circ f = h_2 \circ f$ we have $h_1 = h_2$.

Notice the similarity between these results and those of Exercise 30: in Exercise 30 we showed that a mapping is injective if and only if it is left invertible (and also if and only if it is left cancellable); here we show that a mapping is surjective if and only if it is right invertible (and also if and only if it is right cancellable.)

To prove the results of Exercise 47 we proceed, as in Exercise 30, by showing (a) (1) $\Longrightarrow$ (2), (b) (2) $\Longrightarrow$ (3), (c) $\neg(1) \Longrightarrow \neg(3)$.

For part (a) we apply the Axiom of Choice to the family $(f^{\leftarrow}(b))_{b \in B}$ of non-empty subsets of $A$.

For (b) we form the composition of both sides of the given equation with the mapping $g$.

In (c), if $f$ is not surjective there is an element $b_0$ of $B$ not in the range of $f$. Construct an appropriate set $Y$ and two mappings $h_1$, $h_2$ from $B$ to $Y$ such that $h_1 \circ f = h_2 \circ f$ but $h_1 \neq h_2$.

**Exercise 48.** Let $X$ be a non-empty set of pairwise disjoint non-empty sets. Prove that there is an injection from $X$ to $\bigcup X$.

Apply the Axiom of Choice to the family $(x)_{x \in X}$.

**Exercise 49.** Let $A$ and $B$ be sets, $R$ a relation between $A$ and $B$ with domain $A$. Prove that there exists a mapping $f = ((A, B), F)$ from $A$ to $B$ such that $F \subseteq R$.

Apply the Axiom of Choice to the family $(R^{\rightarrow}(a))_{a \in A}$.

**Exercise 50.** Let $(A_i)_{i \in I}$ and $(A'_i)_{i \in I}$ be families of sets with the same index set $I$. Let $P$ and $P'$ be the products of these families and $(\pi_i)_{i \in I}$, $(\pi'_i)_{i \in I}$ the families of projection mappings. Prove that if for each index $i$ in $I$ there is a mapping $f_i$ from $A_i$ to $A'_i$ then there is a unique mapping $f$ from $P$ to $P'$ such that $\pi'_i \circ f = f_i \circ \pi_i$ for each index $i$ in $I$.

Apply the result of Exercise 45 to the family $(f_i \circ \pi_i)_{i \in I}$.

**Exercise 51.** Prove that if all the mappings $f_i$ in Exercise 50 are injective then the mapping $f$ is injective.

If all the mappings $f_i$ are injective then there are mappings $g_i$ such that $g_i \circ f_i = I_{A_i}$ (see Exercise 30). Apply Exercise 45 to the family $(g_i \circ \pi'_i)_{i \in I}$ to produce a mapping $g$ such that $g \circ f = I_P$.

**Exercise 52.** Prove that if $I = \{1, 2\}$ then there is a bijection from $\prod_{i \in I} X_i$ onto $X_1 \times X_2$.

If $f \in P = \prod_{i \in I} X_i$ then $f$ is a mapping from $I = \{1, 2\}$ to $\bigcup_{i \in I} X_i$

such that $f(1) \in X_1$ and $f(2) \in X_2$. Define a mapping $F$ from $P$ to $X_1 \times X_2$ by setting $F(f) = (f(1), f(2))$ for all elements $f$ of $P$. There is a natural way to define a mapping $G$ from $X_1 \times X_2$ to $P$ by defining for each element $(x_1, x_2)$ of $X_1 \times X_2$ an element of $P$. Prove that $F \circ G$ and $G \circ F$ are the identity mappings of $X_1 \times X_2$ and $P$ respectively.

# Chapter 5

# EQUIVALENCE RELATIONS

Let $X$ be a class and $R$ a relation on $X$. Recall that the diagonal $D_X = \{(x, x) : x \in X\}$. The relation $R$ is said to be

(1) **reflexive** if $D_X \subseteq R$, i.e. if for every $x$ in $X$ we have $(x, x) \in R$;

(2) **irreflexive** if $D_X \cap R = \emptyset$, i.e. if for every $x$ in $X$ we have $(x, x) \notin R$;

(3) **symmetric** if $R = R^{-1}$, i.e. if for every $x$, $y$ in $X$ such that $(x, y) \in R$ we have $(y, x) \in R$;

(4) **antisymmetric** if $R \cap R^{-1} \subseteq D_X$, i.e. if for all $x$, $y$ in $X$ such that $(x, y) \in R$ and $(y, x) \in R$ we have $x = y$.

(5) **transitive** if $R \circ R \subseteq R$, i.e. if for all $x$, $y$, $z$ in $X$ such that $(x, y) \in R$ and $(y, z) \in R$ we have $(x, z) \in R$.

**Exercise 53.** Let $R$ be a relation on a class $X$. Prove that $R \cup R^{-1}$ is the smallest symmetric relation including $R$ and that $R \cap R^{-1}$ is the largest symmetric relation included in $R$.

For the first part show (a) that $R \cup R^{-1}$ is a symmetric relation including $R$ and (b) that every symmetric relation including $R$ includes $R \cup R^{-1}$. Similarly for the second part show (a) that $R \cap R^{-1}$ is a symmetric relation included in $R$ and (b) that every symmetric relation included in $R$ is included in $R \cap R^{-1}$.

**Exercise 54.** Let $R$ be a reflexive and transitive relation on a class $X$. Prove that $R \circ R = R$.

The transitivity of $R$ ensures that $R \circ R \subseteq R$ and the reflexive property is used to prove the reverse inclusion.

Let $X$ be a class; a relation $R$ on $X$ is called an **equivalence relation** if it is reflexive, symmetric and transitive.

**Exercise 55.** Let $X$ be a class, $R$ a relation on $X$. Prove that $R$ is an equivalence relation on $X$ if and only if we have $R \supseteq D_X$ and $R \circ R^{-1} \circ R = R$.

If $R$ is an equivalence relation on $X$ it is clear from the definition that $D_X \subseteq R$. We prove that $R \circ R^{-1} \circ R \subseteq R$ using the symmetry and transitivity of $R$ and the reverse inclusion using the reflexive and transitive properties.

For the converse, the reflexivity of $R$ is immediate. To prove that if $(a, b) \in R$ then $(b, a) \in R$ (and that if $(a, b) \in R$ and $(b, c) \in R$ then $(a, c) \in R$) we have to choose appropriate elements of $R$, $R^{-1}$ and $R$ and use the fact that $R \circ R^{-1} \circ R = R$.

**Exercise 56.** Let $R$ and $S$ be equivalence relations on a class $X$. Prove that $R \circ S$ is an equivalence relation if and only if $R \circ S = S \circ R$ and that when this condition is satisfied $R \circ S$ is the intersection of the class of all equivalence relations on $X$ which include both $R$ and $S$.

**Exercise 57.** Let $R$ be an equivalence relation on a set $E$. Let $A$ be a subset of $R$ such that $\pi_1^{\rightarrow}(A) = E$. Prove that $R \circ A = R$ and that if $S$ is any relation then $(R \cap S) \circ A = R \cap (S \circ A)$.

We have $R \circ A \subseteq R$ for all subsets $A$ of $R$ by the transitivity of $R$.

To show that when $\pi_1^{\rightarrow}(A) = E$ we have $R \subseteq R \circ A$, let $(x, y)$ be any element of $R$; note that there must be an element $t$ of $E$ such that $(x, t) \in A$ and use the symmetry and transitivity of $R$.

In the second part we prove the equality of the sets $(R \cap S) \circ A$ and $R \cap (S \circ A)$ by the usual device of showing that each of them is included in the other.

Let $R$ be an equivalence relation on a set $E$. For each element $a$ of $E$ the **R-class** of $a$ is the set

$$R^{\rightarrow}(a) = \{x : (a, x) \in R\} = \text{ the set of all } R\text{-relatives of } a.$$

We usually denote this set simply by $R(a)$.

Two classes are said to be **disjoint** if their intersection is the empty set $\emptyset$.

**Theorem 1 = Exercise 58.** Let $R$ be an equivalence relation on a set $E$; let $a$ and $b$ be elements of $E$. Then (1) $R(a) = R(b)$ if and only if $(a, b) \in R$; (2) $R(a)$ and $R(b)$ are either equal or disjoint.

Part (1) follows easily from the definition of $R(a)$ and $R(b)$ and appropriate use of the reflexive, symmetric and transitive properties of the relation $R$.

For Part (2) we show that if $R(a)$ and $R(b)$ are not disjoint then they must be equal.

Let $R$ be an equivalence relation on a set $E$. Let $f$ be the mapping from $E$ to $\mathbf{P}(E)$ defined by setting $f(a) = R(a) =$ the $R$-class of $a$ for each element $a$ of $E$. The range of $f$ is denoted by $E/R$ and is called the **quotient set of $E$ by the equivalence relation $R$**. The mapping $\eta$ from $E$ to $E/R$ defined by setting $\eta(a) = R(a)$ for each element $a$ of $E$ is surjective and is called the **canonical surjection** from $E$ onto $E/R$.

**Exercise 59.** Let $(A_i)_{i\in I}$ be a family of sets with index set $I$. For each index $i$ in $I$ let $R_i$ be an equivalence relation on $A_i$. Let $P$ be the product of the family $(A_i)_{i\in I}$ and $(\pi_i)_{i\in I}$ the family of projection mappings from $P$. Let $R$ be the relation on $P$ defined by setting $(f, g) \in R$ if and only if $(\pi_i(f), \pi_i(g)) \in R_i$ for all indices $i$ in $I$. Prove that $R$ is an equivalence relation on $P$ and that there is a "natural" bijection $F$ from $P/R$ onto the product $P'$ of the family $(A_i/R_i)_{i\in I}$.

The proof that $R$ is an equivalence relation is straightforward.

The canonical surjections $\eta$ from $P$ onto $P/R$ and $\eta_i$ from $A_i$ to $A_i/R_i$ for each index $i$ in $I$ have right inverses $\nu$ and $\nu_i$ $(i \in I)$. To define a mapping $F$ from $P/R$ to $P' = \prod_{i\in I} A_i/R_i$, consider each element $X$ of $P/R$ and look at $\nu(X)$ in $P$; $\nu(X)$ provides for each index $i$ in $I$ an element of $A_i$ and consequently, on application of $\eta_i$, an element of $A_i/R_i$. These elements serve to determine an element of $P'$.

The proof that the mapping $F$ so defined is bijective rests on the definition of the equivalence relation $R$ and uses the mappings $\nu_i$.

Let $A$ and $B$ be sets, $f$ a mapping from $A$ to $B$ with graph $F$. Let $R_f$ be the relation on $A$ defined by setting $(x, y) \in R_f$ if and only if $f(x) = f(y)$. Then $R_f$ is an equivalence relation on $A$ which we call the **equivalence relation associated with** $f$. Exercises 60 and 61 are straightforward applications of this definition.

**Exercise 60.** With the notation just described, prove that $R_f = F^{-1} \circ F$.

**Exercise 61.** Let $R$ be an equivalence relation on a set $E$; let $\eta$ be the canonical surjection from $E$ onto the quotient set $E/R$. Prove that $R$ is the equivalence relation associated with $\eta$.

Let $R$ be an equivalence relation on a set $E$. A subset $A$ of $E$ is said to be **saturated with respect to** $R$ if for every element $a$ of $A$ the $R$-class of $a$ is included in $A$. Thus $A$ is saturated with respect to $R$ if and only if $A$ is a union of $R$-classes.

**Exercise 62.** Let $R$ be an equivalence relation on a set $E$, $\eta$ the canonical surjection from $E$ onto $E/R$, $A$ a subset of $E$. Prove that $A$ is saturated with respect to $R$ if and only if $\eta^{\leftarrow}(\eta^{\rightarrow}(A)) = A$.

If $A$ is a subset of $E$ such that $\eta^{\leftarrow}(\eta^{\rightarrow}(A)) = A$ let $a$ be any element of $A$, $b$ any element of the $R$-class of $a$; then show that $b \in A$.

To prove the converse suppose $A$ is saturated. We have only to prove that $\eta^{\leftarrow}(\eta^{\rightarrow}(A)) \subseteq A$ since the reverse inclusion holds for all subsets $A$ of $E$ (see Exercise 33(12)). So let $a$ be any element of $\eta^{\leftarrow}(\eta^{\rightarrow}(A))$ and deduce that $a \in R(a_1)$ for some element $a_1$ of $A$.

Let $R$ be an equivalence relation on a set $E$; let $B$ be any subset of $E$. Then $\eta^{\leftarrow}(\eta^{\rightarrow}(B))$ is the smallest saturated subset of $E$ which includes $B$. We call it the **saturation** of $B$ with respect to $R$.

**Exercise 63.** Let $R$ be an equivalence relation on a set $E$, $\eta$ the canonical surjection from $E$ onto $E/R$. Let $f$ be a mapping from $E$ to a set $F$. Prove that there is a mapping $h$ from $E/R$ to $F$ such that $f = h \circ \eta$ if and only if for all $(x, y)$ in $R$ we have $f(x) = f(y)$.

The "only if" part is clear.

To prove the "if" part, let $\nu$ be a mapping from $E/R$ to $E$ such that $\eta \circ \nu = I_{E/R}$ and note that for each element $x$ of $E$ we have $(\nu(\eta(x)), x) \in R$.

**Exercise 64.** Let $f$ be a mapping from a set $E$ to a set $F$, $R$ and $S$ equivalence relations on $E$ and $F$ respectively, $\eta_R$ and $\eta_S$ the corresponding canonical surjections. Prove that there is a mapping $h$ from $E/R$ to $F/S$ such that $h \circ \eta_R = \eta_S \circ f$ if and only if for all $(x, y)$ in $R$ we have $(f(x), f(y)) \in S$.

Again the "only if" part is straightforward.

To prove the "if" part, notice that whenever $(x, y) \in R$ we have $(\eta_S \circ f)(x) = (\eta_S \circ f)(y)$ and use Exercise 63.

**Exercise 65.** Let $R$ and $S$ be equivalence relations on a set $E$ such that $S \subseteq R$. Show how to define a "natural" equivalence relation $R/S$ on $E/S$ in such a way that there exists a bijection from $E/R$ onto $(E/S)/(R/S)$.

Let $\eta_R$ and $\eta_S$ be the canonical surjections from $E$ onto $E/R$ and $E/S$ respectively; let $\nu_R$ and $\nu_S$ be mappings from $E/R$ and $E/S$ to $E$ such that $\eta_R \circ \nu_R$ and $\eta_S \circ \nu_S$ are the identity mappings of $E/R$ and $E/S$ respectively. Define a relation $R/S$ on $E/S$ by setting $(X, Y) \in R/S$ if and only if $(\nu_S(X), \nu_S(Y)) \in R$. Clearly $R/S$ is an equivalence relation; let $\eta_{R/S}$ be the canonical surjection of $E/S$ onto $(E/S)/(R/S)$. Prove that $f = \eta_{R/S} \circ \eta_S \circ \nu_R$ is the required bijection.

# Chapter 6

# ORDER RELATIONS

Let $X$ be a class, $R$ a relation on $X$. Then $R$ is said to be an **order relation**, **ordering** or **order** on $X$ if it is reflexive, antisymmetric and transitive. So $R$ is an order on $X$ if and only if $D_X \subseteq R$, $R \cap R^{-1} \subseteq D_X$ and $R \circ R \subseteq R$. An **ordered set** is an ordered pair $(E, R)$ such that $E$ is a set and $R$ is an order on $E$. If the order $R$ is clear from the context we may refer to the ordered set $(E, R)$ simply as "the ordered set $E$". Some authors prefer to use the term"partial order" for what we have called an order and instead of ordered sets speak of "partially ordered sets" or "posets".

Exercises 66 and 67 follow at once from the definition of order.

**Exercise 66.** Prove that if $R$ is an order on a class $X$ then so is the inverse relation $R^{-1}$.

**Exercise 67.** Prove that a relation $R$ on a class $X$ is an order if and only if $R \circ R = R$ and $R \cap R^{-1} = D_X$.

If $R$ is an order on a class $X$ we usually write $a \leq b$ instead of $(a, b) \in R$; we read this as "$a$ is less than or equal to $b$" or "$b$ is greater than or equal to $a$"; we also write $b \geq a$ synonymously with $a \leq b$. (We occasionally write $a \leq_R b$ and $a \geq_R b$ if it is necessary to clarify which order $R$ is in use.) Using this notation we see that a relation $R$ is an order on a class $X$ if

(1) for all elements $x$ of $X$ we have $x \leq x$;

(2) for all elements $x$, $y$ of $X$ such that $x \leq y$ and $y \leq x$ we have $x = y$;

(3) for all elements $x$, $y$, $z$ of $X$ such that $x \leq y$ and $y \leq z$ we have $x \leq z$.

If $R$ is an order on a class $X$ and we write $a \leq b$ instead of $(a, b) \in R$, then we also write $a < b$ and $b > a$ (read "$a$ is less than $b$", "$b$ is greater than $a$") or, if necessary, $a <_R b$ and $b >_R a$ as abbreviations for "$(a, b) \in R$ and $a \neq b$". Let $R$ be an order on a class $X$ and write $x \leq y$ instead of $(x, y) \in R$. Then an element $a$ of $X$ is called

(1) an $R$-**least** or $R$-**smallest** or $R$-**first** element of $X$ if $a \leq x$ for all elements $x$ of $X$;

(2) an $R$-**greatest** or $R$-**largest** or $R$-**last** element of $X$ if $x \leq a$ for all elements $x$ of $X$;

(3) an $R$-**minimal** element of $X$ if for every element $x$ of $X$ such that $x \leq a$ we have in fact $x = a$;

(4) an $R$-**maximal** element of $X$ if for every element $x$ of $X$ such that $x \geq a$ we have in fact $x = a$.

If $R$ is an order on a class $X$ there can be at most one $R$-least element of $X$; if $X$ has an $R$-least element it is also the unique $R$-minimal element of $X$. If $X$ has no $R$-least element there may be many $R$-minimal elements or none at all. Similar remarks apply to $R$-greatest and $R$-maximal elements.

Let $X$ be a class, $R$ an order on $X$ and $A$ a subclass of $X$. Let $b$ be an element of $X$. Then $b$ is called

(1) an $R$-**lower bound** for $A$ if for every element $a$ of $A$ we have $b \leq a$;

(2) an $R$-**upper bound** for $A$ if for every element $a$ of $A$ we have $a \leq b$;

(3) an $R$-**greatest lower bound** for $A$ if $b$ is an $R$-greatest element of the set of $R$-lower bounds for $A$;

(4) an $R$-**least upper bound** for $A$ if $b$ is an $R$-least element of the set of all $R$-upper bounds for $A$.

A subclass $A$ of $X$ is said to be **bounded above** (**bounded below**) if it has at least one $R$-upper bound ($R$-lower bound). A subclass which is bounded above need not have an R-least upper bound, but it cannot have more than one. Similar remarks apply to subclasses which are bounded below.

If $A$ has an $R$-least upper bound $b$ we write $R$-sup $A = b$; if $A$ has an $R$-greatest lower bound $c$ we write $R$-inf $A = c$ (*sup* and *inf* are derived from the Latin words *supremum* and *infimum*, meaning "greatest" and "least" respectively and are often read in full).

**Exercise 68.** Let $A$ and $B$ be subclasses of a class $X$; let $R$ be an order on $X$. If $A \subseteq B$ and $A$ and $B$ both have $R$-least upper and $R$-greatest lower bounds, prove that we have $R$-sup $A \leq R$-sup $B$ and $R$-inf $A \geq R$-inf $B$.

Show that $R$-sup $B$ and $R$-inf $B$ are respectively $R$-upper and $R$-lower bounds for $A$.

**Exercise 69.** Let $R$ be an order on a class $X$. Let $A$ be a subclass of $X$ such that $A = \bigcup_{i \in I} A_i$ where $(A_i)_{i \in I}$ is a family of subclasses of $X$ with index set $I$. Suppose each of the subclasses $A_i$ has an $R$-least upper bound $b_i$; let $B$ be the set of all the bounds $b_i$. Prove that $A$ has an $R$-least upper bound if and only if $B$ has an $R$-least upper bound and that, when this condition is satisfied, $R$-sup $A = R$-sup $B$.

Show that every $R$-upper bound for $B$ is an $R$-upper bound for $A$ and conversely.

Let $(E, R)$ be an ordered set. Then $(E, R)$ is said to be

(1) a **directed set** if every two-element subset of $E$ is bounded above;

(2) a **lattice** if every two-element subset of $E$ has both an $R$-least upper bound and an $R$-greatest lower bound;

(3) a **totally ordered set, linearly ordered set**, or a **chain** (and $R$ is called a **total order** or a **linear order**) if for every pair of elements $x$, $y$ of $E$ we have either $(x, y) \in R$ or $(y, x) \in R$;

(4) an **inductively ordered set** if every subset of $E$ which is totally ordered by $R$ has an $R$-upper bound;

(5) a **well-ordered set** (and $R$ is called a **well-ordering**) if every non-empty subset of $E$ has an $R$-least element.

**Exercise 70.** Let $V$ be a vector space over a field $F$; let $E$ be the set of all subspaces of $V$. Prove that $(E, \subseteq)$ is a lattice.

If $S$ and $T$ are subspaces of $V$ we must find the largest subspace of $V$ included in $S$ and $T$ and the smallest subspace of $V$ which includes $S$ and $T$.

**Exercise 71.** Let $(E, R)$ be a lattice; let $I$ and $J$ be finite sets, $(x_{ij})$ a family of elements of $E$ indexed by $I \times J$. Prove that

$$(\sup_{j \in J}(\inf_{i \in I} x_{ij}), \inf_{i \in I}(\sup_{j \in J} x_{ij})) \in R.$$

Start by noticing that for every index $i_0$ in $I$ and every index $j_0$ in $J$ we have

$$\inf_{i \in I} x_{ij_0} \leq x_{i_0 j_0} \leq \sup_{j \in J} x_{i_0 j}.$$

**Exercise 72.** Let $E$ be a set, $Q$ the set of all orders on $E$, $R$ the inclusion relation on $\mathbf{P}(E \times E)$. Prove that $T$ is an $R$-maximal element of $Q$ if and only if $T$ is a total order on $E$.

It is easy to show that any order $S$ on $E$ which includes a total order $T$ must coincide with $T$.

To prove the converse let $T$ be an order on $E$ which is not a total order; then there are elements $a$ and $b$ of $E$ such that neither $(a, b)$ nor $(b, a)$ belongs to $T$. Show that

$$T' = T \cup \{(x, y) : ((x, y) \in E \times E) \wedge ((x, a) \in T) \wedge ((b, y) \in T)\}$$

is an order on $E$ which properly includes $T$, hence showing that $T$ is not $(\subseteq)$-maximal in $Q$.

**Exercise 73.** Let $E$ be a set, $S$ a non-empty set of orders on $E$ such that $S$ is totally ordered by the inclusion relation on $\mathbf{P}(E \times E)$. Prove that $\bigcup S$ is an order on $E$.

The checking of the order conditions for $\bigcup S$ is straightforward as soon as we recall that if $R_1$ and $R_2$ are orders in $S$ we have either $R_1 \subseteq R_2$ or $R_2 \subseteq R_1$.

Let $(E, R)$ and $(E', R')$ be ordered sets; let $f$ be a mapping from $E$ to $E'$. Then $f$ is said to be

(1) an **isomorphism** if it is a bijection and $(x, y) \in R$ if and only if $(f(x), f(y)) \in R'$;

(2) **increasing** if for all elements $x$, $y$ of $E$ such that $(x, y) \in R$ we have $(f(x), f(y)) \in R'$;

(3) **decreasing** if for all elements $x$, $y$ of $E$ such that $(x, y) \in R$ we have $(f(y), f(x)) \in R'$;

(4) **monotone** if it is either increasing or decreasing;

(5) **strictly increasing** if for all elements $x$, $y$ of $E$ such that $x \neq y$ and $(x, y) \in R$ we have $f(x) \neq f(y)$ and $(f(x), f(y)) \in R'$;

(5) **strictly decreasing** if for all elements $x$,$y$ of $E$ such that $x \neq y$ and $(x, y) \in R$ we have $f(x) \neq f(y)$ and $(f(y), f(x)) \in R'$;

If it is necessary to emphasize the order relations $R$, $R'$ we may talk of $(R, R')$-isomorphisms, $(R, R')$-increasing mappings and so on.

If there exists an isomorphism from $E$ to $E'$ then $(E, R)$ and $(E', R')$ are said to be **isomorphic**. It is easy to verify that if $f$ is an isomorphism with respect to $R$ and $R'$ then $f^{-1}$ is an isomorphism from $E'$ to $E$ with respect to $R'$ and $R$.

**Exercise 74.** Let $(E, R)$ and $(E', R')$ be ordered sets. Let $f$ be a decreasing mapping from $E$ to $E'$ and $g$ a decreasing mapping from $E'$ to $E$. Suppose that $(x, g(f(x))) \in R$ for all elements $x$ of $E$ and $(x', f(g(x'))) \in R'$ for all elements $x'$ of $E'$. Prove that $f \circ g \circ f = f$ and $g \circ f \circ g = g$.

Let $x$ be any element of $E$. Apply the fact that $f$ is decreasing to the first condition and take $x' = f(x)$ in the second condition.

**Exercise 75.** Let $(E, R)$ and $(E', R')$ be lattices, $f$ a mapping from $E$ to $E'$. Prove that $f$ is increasing if and only if for all elements $x$, $y$ of $E$ we have $(f(R\text{-inf}\{x, y\}), R'\text{-inf}\{f(x), f(y)\}) \in R'$.

If $f$ is increasing and $z = R\text{-inf}\ \{x, y\}$ notice that $(f(z), f(x)) \in R$ and $(f(z), f(y)) \in R$.

If the condition holds and $(x, y) \in R$ then make use of the fact that $x = R\text{-inf}\ \{x, y\}$.

# Chapter 7

# WELL-ORDERING

Let $X$ be a class, $R$ an order on $X$. We recall that $R$ is a total order on $X$ if for all elements $x$, $y$ of $X$ we have either $(x, y) \in R$ or $(y, x) \in R$; $R$ is called a **well-ordering** on $X$ if every non-empty subclass of $X$ has an $R$-least element.

**Theorem 1 = Exercise 76.** Every well-ordering on a class $X$ is a total order on $X$.

This is immediate when we apply the well-ordering condition to the subsets of $X$ which are unordered pairs of elements.

Let $R$ be an order on a class $X$. A subclass $S$ of $X$ is called an $R$-**segment** if for every element $s$ of $S$ and every element $x$ of $X$ such that $(x, s) \in R$ we have $x \in S$. If $a$ is any element of $X$ then the set

$$S_{X,R}(a) = \{x : (x \in X) \wedge ((x, a) \in R) \wedge (x \neq a)\} = \{x \in X : x < a\}$$

is an $R$-segment of $X$ which we call the **initial $R$-segment determined by** $a$.

If $R$ is an order on a class $X$ and $A$ is a subclass of $X$ we write $R|A = R \cap (A \times A)$ and call it the **restriction** of $R$ to $A$.

**Exercise 77.** Prove that $R|A$ is an order on $A$. Show that $R|A$ is a total order if $R$ is a total order and a well-ordering if $R$ is a well-ordering.

This involves only careful checking of the defining conditions.

**Exercise 78.** Let $R$ be an order on a class $X$. Show that for every element $x$ of $X$ we have $S_{X,R}(x) \subset X$. Prove also

(1) if $A$ is a subclass of $X$ and $a \in A$ then $S_{A,R|A}(a) \subseteq S_{X,R}(a)$;

(2) if $a$ and $b$ are elements of $X$ such that $(a,b) \in R$ then we have $S_{X,R}(a) \subseteq S_{X,R}(b)$;

(3) if $a$ is an element of $X$ and $S = S_{X,R}(a)$ then for every element $x$ of $S$ we have $S_{S,R|S}(x) = S_{X,R}(x)$.

The arguments here are all straightforward.

**Theorem 2 = Exercise 79.** Let $R$ be a well-ordering on a class $X$. Then every proper $R$-segment, i.e. every $R$-segment which is a proper subset of $X$, is an initial segment.

Consider the (non-empty) complement of $X$, which must have an $R$-least element.

**Exercise 80.** Let $R$ be a well-ordering on a set $E$. Show that the set of $R$-segments of $E$ is well-ordered by the inclusion relation.

It is clearly sufficient to prove that the set of proper $R$-segments is well-ordered; these are all initial segments. So a non-empty subset of proper segments gives rise to a non-empty subset of $E$.

**Theorem 3 = Exercise 81** (Principle of Transfinite Induction). Let $R$ be a well-ordering on a class $X$, $A$ a subclass of $X$. If for every element $x$ of $X$ we have the implication $S_{X,R}(x) \subseteq A \Longrightarrow x \in A$ then $A = X$.

Suppose, to the contrary, that $X \neq A$; then the complement of $A$ is non-empty and so has an $R$-least element, $b$ say. Show that $S_{X,R}(b) \subseteq A$ and deduce a contradiction.

**Theorem 4 = Exercise 82.** Let $R$ and $R'$ be relations of well-ordering on classes $E$ and $E'$ respectively. Let $F$ and $G$ be functional relations between $E$ and $E'$ such that for all elements $a$, $b$ of $E$ such that $(a,b) \in R$ we have $(F(a), F(b)) \in R'$ and $(G(a), G(b)) \in R'$. If (1) $F^{\rightarrow}(E)$ is a segment of $E'$ and (2) for all elements $a$, $b$ of $E$ such that $a \neq b$ we have $G(a) \neq G(b)$ then $(F(x), G(x)) \in R'$ for all elements $x$ of $E$.

Suppose, to the contrary, that the subclass

$$\{t : (t \in E) \wedge ((F(t), G(t)) \notin R')\}$$

is non-empty; let $a$ be its $R$-least element. Show that $G(a) \in F^{\rightarrow}(E)$ and deduce the contradiction that $(G(a), G(a)) \notin R'$.

The next two exercises are immediate consequences of Theorem 4.

**Exercise 83.** Let $(E, R)$ and $(E', R')$ be well-ordered sets, $f$ and $g$ increasing mappings from $E$ to $E'$ such that (1) $f^{\rightarrow}(E)$ is a segment of $E'$ and (2) $g$ is strictly increasing. Show that $f(x) \leq g(x)$ for all elements $x$ of $E$.

Suppose, to the contrary, that the set $\{t : (t \in E) \wedge (f(t) > g(t))\}$ is non-empty; let $a$ be its $R$-least element. Show that $g(a) \in f^{\rightarrow}(E)$ and deduce the contradiction that $g(a) < g(a)$.

**Exercise 84.** Let $(E, R)$ and $(E', R')$ be well-ordered sets. There is at most one isomorphism from $E$ onto $E'$.

**Exercise 85.** Let $(E, R)$ be a well-ordered set. There exists no isomorphism from $E$ onto an initial segment of $E$.

Suppose there were an isomorphism $f$ from $E$ onto the initial segment $S_{E,R}(a)$. Apply Theorem 4 to $f$ and $I_R$ and deduce the contradiction that $a < a$.

**Theorem 5 = Exercise 86.** Let $(E, R)$ and $(E', R')$ be well-ordered sets. Then either (1) there exists a unique isomorphism from $E$ onto $E'$ or (2) there exists a unique isomorphism from $E$ onto an initial segment of $E'$ or (3) there exists a unique isomorphism from $E'$ onto an initial segment of $E$.

If we can prove the existence of an isomorphism as described then Exercise 84 takes care of its uniqueness.

Let $A$ be the subset of $E$ consisting of those elements $x$ such that there is an isomorphism from the initial segment $S_{E,R}(x)$ onto an initial segment $S_{E',R'}(x')$ of $E'$. Define the mapping $f$ from $A$ to $E'$ by setting, for each element $x$ of $A$, $f(x)$ = the corresponding element $x'$; let $A' = f^{\rightarrow}(A)$. Clearly $f$ is an isomorphism from $A$ onto $A'$. Consider the four cases (1) $A = E$ and $A' = E'$; (2) $A = E$ and $A' \subset E'$; (3) $A \subset E$ and $A' = E'$; (4) $A \subset E$ and $A' \subset E'$. In case (1) we are done. In case (2) show that $A'$ is an initial segment of $E'$. Case (3) is the same as case (2) for the inverse isomorphism $f^{-1}$. Show that case (4) cannot occur.

# Chapter 8

# ORDINALS

**Axiom 8 (Axiom of Foundation or Regularity).** For every non-empty class $X$ there exists an element $x$ of $X$ such that $X \cap x = \emptyset$.

It follows from this axiom that if $a$ is any set then $a \notin a$. For if $a \in a$ then we have $a \in a \cap \{a\}$; but $a$ is the only member of $\{a\}$, so there is no member of $\{a\}$ disjoint from it, contradicting the Axiom of Foundation. It follows that the Russell class $R$ (see p. 11) is equal to the universe $V$.

Let $E$ be the class of all ordered pairs $(x, y)$ such that $x \in y$ or $x = y$. For each class $X$ let $E \,|\, X = E \cap (X \times X)$.

A class $X$ is said to be an **ordinal** if (1) $E \,|\, X$ is a well-ordering on $X$; (2) every element of $X$ is a subset of $X$. A set $X$ which is an ordinal is called an **ordinal number.** We denote the class of all ordinal numbers by $On$; we usually use Greek letters to denote ordinal numbers.

The next exercise, which involves only straightforward checking, gives simple examples of ordinal numbers.

**Exercise 87.** Assuming that the null class $\emptyset$ is a set (see Exercise 100), show that

$$0 = \emptyset; 1 = \{0\}; 2 = \{0, 1\}; 3 = \{0, 1, 2\} \ldots$$

are ordinal numbers.

**Theorem 1 = Exercise 88.** If $X$ and $Y$ are ordinals and there is an isomorphism $f$ from $X$ onto $Y$ (relative to the orders $E \mid X$ and $E \mid Y$) then $X = Y$.

Let $A = \{x : (x \in X) \wedge (f(x) = x)\}$ and show that $A = X$ by the usual device of assuming the contrary and considering the $E$-least element $u$ of its complement; obtain a contradiction by showing that $u = f(u)$.

**Theorem 2 = Exercise 89.** Let $X$ be an ordinal. If $x$ is an element of $X$ then $x$ is an ordinal number and $x = S_{X,E \mid X}(x)$.

It is clear that $x$ is a set, well-ordered by $E \mid x$. To show that each element $u$ of $x$ is a subset of $x$ let $v$ be any element of $u$ and use the transitivity of $E \mid X$. The proof that $x = S_{X,E \mid X}(x)$ is achieved by the usual procedure of establishing two inclusion relations.

**Theorem 3 = Exercise 90.** $On$ is an ordinal.

It is straightforward to show that $E \mid On$ is an order relation on $On$. To show that it is a well-ordering let $x$ be any non-empty subset of $On$ and $\alpha$ any element of $x$; if $\alpha$ is not the $(E \mid On)$-least element of $x$ consider the non-empty set $\alpha \cap x$. It is clear that if $\alpha \in On$ then $\alpha \subseteq On$.

**Corollary 1 = Exercise 91.** $On$ is a proper class.

Use Exercise 90 to show that if $On$ were a set we would have a contradiction to Axiom 8.

**Corollary 2 = Exercise 92.** For every ordinal number $\alpha$ we have $\alpha = S_{On,E \mid On}(\alpha)$.

Use Exercise 89.

Let $\beta$ be an ordinal number; the **successor** of $\beta$ is the set $\beta^+ = \beta \cup \{\beta\}$.

**Exercise 93.** Prove that for every ordinal number $\alpha$ we have $\bigcup \alpha^+ = \alpha$.

This yields to the usual two-part strategy for proving two sets equal.

**Exercise 94.** Prove that if $\beta$ is an ordinal number then so is $\beta^+$.

Use the Pairing and Union Axioms; show that $E \mid \beta^+$ is a well-ordering on $\beta^+$ and that every element $x$ of $\beta^+$ is a subset (distinguish the cases where $x \in \beta$ and $x = \beta$).

In pre-axiomatic set theory, where it was assumed without question that every open sentence determined a set, the class $On$ (determined by the open sentence "$x$ is an ordinal number ") was taken to be a set. Thus $On$, being an ordinal, was taken to be an ordinal number. Hence (see Exercise 94) its successor $On^+$ was also an ordinal number and hence $On^+ \in On$, which is not the case. It was this contradiction which was known as Burali-Forti's paradox.

An ordinal number $\alpha$ is said to be a **limit ordinal** if it is non-empty and there is no ordinal number $\beta$ such that $\alpha = \beta^+$. An ordinal number $\alpha$ is called a **non-limit ordinal** if either $\alpha$ is empty or there is an ordinal number $\beta$ such that $\alpha = \beta^+$.

**Exercise 95.** Let $\alpha$ be a non-empty ordinal number. Prove that $\alpha$ is a limit ordinal if and only if for every element $\beta$ of $\alpha$ its successor $\beta^+$ is also an element of $\alpha$.

The "if" part is easily proved by contradiction. For the "only if" part let $\beta$ be any element of a limit ordinal $\alpha$ and examine the three possibilities (1) $\beta^+ \in \alpha$, (2) $\beta^+ = \alpha$ and (3) $\alpha \in \beta^+$, ruling out the second and third.

**Theorem 4 = Exercise 96.** For every well-ordered set $(x, R)$ there is a unique ordinal number $\alpha$ such that $(\alpha, E \mid \alpha)$ is isomorphic to $(x, R)$.

If $X$ and $Y$ are classes ordered by $R$ and $S$ respectively we amend the meaning of the term *isomorphism* as follows: an isomorphism $F$ from $X$ to $Y$ is a functional relation between $X$ and $Y$ such that Domain $F = X$, Range $F = Y$, $F^{-1}$ is functional and whenever

$(x_1, x_2) \in R$, $(x_1, y_1) \in F$ and $(x_2, y_2) \in F$ we have $(y_1, y_2) \in S$. An easy adaptation of Exercise 86 shows that if $X$ and $Y$ are classes well-ordered by $R$ and $S$ then (using the new interpretation of "isomorphism") either (1) there exists a unique isomorphism from $X$ to $Y$ or (2) there exists a unique isomorphism from $X$ to an initial segment of $Y$ or (3) there exists a unique isomorphism from $Y$ to an initial segment of $X$.

Apply this result to $x$, well-ordered by $R$ and $On$, well-ordered by $E$, ruling out possibilities (1) and (3) by using the Replacement Axiom.

The unique ordinal number $\alpha$ such that $(\alpha, E \mid \alpha)$ is isomorphic to $(x, R)$ is called the **ordinal number of** the well-ordered set $(x, R)$ and is sometimes denoted by $\mathrm{Ord}(x, R)$ or $\mathrm{Ord}(x)$ if the order $R$ is clear from the context.

**Exercise 97.** Let $(x, R)$ and $(y, S)$ be well-ordered sets. Prove that $\mathrm{Ord}(x, R) = \mathrm{Ord}(y, S)$ if and only if there exists an $(R, S)$-isomorphism from $x$ to $y$.

**Exercise 98.** Prove that there exists a relation of well-ordering on a set $x$ if and only if $x$ is equinumerous with an ordinal number.

This follows at once from Exercise 96.

**Theorem 5 = Exercise 99** (Transfinite Recursion Theorem). Let $F$ be a functional relation whose domain is the class of all sets. Then there exists a unique functional relation $G$ with domain $On$ such that for every ordinal number $\alpha$ we have $G(\alpha) = F(G^{\rightarrow}(\alpha))$.

Informally we understand the equation $G(\alpha) = F(G^{\rightarrow}(\alpha))$ to mean that the value of the functional relation $G$ for the ordinal number $\alpha$ is obtained by applying the functional relation $F$ to the set of values of $G$ for all the ordinal numbers $\beta$ in $\alpha = S_{On, E \mid On}(\alpha)$, i.e. all the ordinal numbers $\beta$ which are $E$-less than $\alpha$.

Let $M$ be the class of all functional relations $H$ such that $\mathrm{Dom}\, H \subseteq On$ and for all ordinal numbers $\alpha$ in $\mathrm{Dom}\, H$ we have $\alpha \subseteq \mathrm{Dom}\, H$ and $H(\alpha) = F(H^{\rightarrow}(\alpha))$. Apply the ideas of Exercise 43 to the family $(H)_{H \in M}$.

# Chapter 9

# NATURAL NUMBERS

**Axiom 9 (Axiom of Infinity).** There exists a set $y$ such that $\emptyset \in y$ and whenever $x \in y$ then $x^+ \in y$.

**Exercise 100.** Show that the null class is a set.

This is an immediate consequence of the Axiom of Infinity.

A set $n$ is called a **natural number** if (1) $n$ is an ordinal number and (2) every non-empty subset of $n$ has an $E$-greatest element. The class of natural numbers is denoted by $\omega$. A set is said to be **finite** if it is equipotent to (equinumerous with) a natural number.

**Exercise 101.** Show that every element of a natural number is a natural number.

Use Exercise 89 and the fact that every element of an ordinal is a subset.

**Theorem 1 = Exercise 102.** If $\alpha$ is an ordinal and $\beta$ is an $E$-greatest member of $\alpha$ then $\alpha = \beta^+$.

It is clear that $\beta \cup \{\beta\} \subseteq \alpha$; if $x \in \alpha$ then $(x, \beta) \in E$, whence $x \in \beta$ or $x = \beta$.

**Theorem 2 = Exercise 103.** If $n$ is a natural number then $n^+$ is a natural number.

Use Exercise 94.

**Theorem 3 = Exercise 104.** $\emptyset$ is a natural number; $\emptyset$ is not the successor of any natural number.

Show that $\emptyset$ satisfies the defining properties of natural numbers and note that successors are by definition non-empty.

**Theorem 4 = Exercise 105.** If $m$ and $n$ are natural numbers such that $m^+ = n^+$ then $m = n$.

Use Exercise 93.

**Theorem 5 = Exercise 106.** If $x$ is a subclass of $\omega$ such that (1) $\emptyset \in x$ and (2) $u^+ \in x$ whenever $u \in x$, then $x = \omega$.

Suppose $x \neq \omega$, so that $\omega \sim x$ is non-empty and hence has an $E$-least element $\alpha$, which cannot be $\emptyset$ and so must have an $E$-greatest element $\beta$; by Theorem 1 we have $\alpha = \beta^+$ and this leads to a contradiction.

Theorems 2, 3, 4 and 5 are the **Peano postulates for the natural numbers.**

**Theorem 6 = Exercise 107.** $\omega$ is an ordinal number.

It is easy to show that $\omega$ is an ordinal. To show that it is a set let $y$ be the set whose existence is asserted by the Axiom of Infinity; apply Theorem 5 to the class $y \cap \omega$.

**Theorem 7 = Exercise 108** (Principle of Mathematical Induction). Let $P(n)$ be an open sentence in which $n$ occurs as a free variable and for which we have

(1) $P(\emptyset)$ and

(2) $(\forall k)((k \in \omega) \Longrightarrow (P(k) \Longrightarrow P(k^+)))$.

Prove that

$(\forall n)((n \in \omega) \Longrightarrow P(n))$.

Apply Exercise 106 to the class $x = \{n : (n \in \omega) \wedge P(n)\}$.

**Theorem 8 = Exercise 109.** Let $E$ be a set; if $a$ is an element of $E$ and $f$ a mapping from $E$ to itself then there exists a unique mapping $g$ from $\omega$ to $E$ such that $g(0) = a$ and $g(n^+) = f(g(n))$ for every natural number $n$.

The uniqueness of the mapping $g$ follows easily by mathematical induction. To prove the existence of $g$ we adapt the ideas of the proof of Exercise 99.

# Chapter 10

# EQUIVALENTS OF THE AXIOM OF CHOICE

**Exercise 110.** Suppose that for every set $x$ there exists a well-ordering on $x$. Deduce the Axiom of Choice.

Let $(x_i)_{i \in I}$ be a family of non-empty sets. If $I = \emptyset$ then $\prod_{i \in I} x_i$ is non-empty by its definition. If $I$ is non-empty let $x = \bigcup_{i \in I} x_i$; use a well-ordering on $x$ to produce an element of $\prod_{i \in I} x_i$.

**Theorem 1 (Zermelo's Theorem) = Exercise 111.** If the Axiom of Choice holds then for every set $x$ there exists a well-ordering on $x$.

If $x = \emptyset$ then $\emptyset$ is a well-ordering on $x$.

So suppose $x$ is non-empty. By the Axiom of Choice we have a mapping $f_1$ from the set of non-empty subsets of $x$ to $x$ such that $f_1(y) \in y$ for every non-empty subset $y$ of $x$. Extend $f_1$ to a mapping $f$ from $\mathbf{P}(x)$ to $x \cup \{u\}$ where $u \notin x$ by setting $f(y) = f_1(y)$ if $y \neq \emptyset$ and $f(\emptyset) = u$. Use the Transfinite Recursion Theorem (Exercise 99) to produce a functional relation $G$ with domain $On$ such that $G(\alpha) = f(x \sim G^{\rightarrow}(\alpha))$ for all ordinal numbers $\alpha$.

Let $Y$ be the subclass of $On$ consisting of ordinal numbers $\alpha$ for which $G(\alpha) \neq u$. Show that $Y$ is a segment of $On$ and that $G \mid Y$ is an injection. Deduce that $Y$ must be a *proper* segment of $On$, so an ordinal number, and then that $((Y, x), G \mid Y)$ is a bijection.

Let $P(X)$ be a condition involving the variable $X$. Then $P(X)$ is said to be a **property of finite character** if for all $x$ we have $P(x)$ if and only if $P(y)$ holds for all finite subsets $y$ of $x$.

The following two exercises are straightforward, though of course some ingenuity is required in Exercise 113 when counterexamples are asked for.

**Exercise 112.** Show that each of the following properties $P(X)$ is a property of finite character:

(1) $(X, R \cap (X \times X))$ is a totally ordered set (where $R$ is a given relation);

(2) $u \notin X$ (where $u$ is a given set);

(3) the elements of $X$ are pairwise disjoint sets.

**Exercise 113.** Let $P(X)$ be a property of finite character. For each of the following, if it is a theorem, prove it, and if not, give a counterexample:

(1) for all sets $x$ and $y$ such that $P(x)$ and $P(y)$ we have $P(x \cup y)$.

(2) for all sets $x$ and $y$ such that $P(x)$ and $P(y)$ we have $P(x \cap y)$.

The following three propositions are known as **maximal principles**:

**1. Hausdorff's maximal principle.** Let $(x, R)$ be an ordered set. Let

$$T = \{y : y \in \mathbf{P}(x) \text{ and } R \cap (y \times y) \text{ is a relation of total order on } y\}.$$

Then $T$ has a $(\subseteq)$-maximal element.

**2. Zorn's Lemma.** Let $(x, R)$ be an inductively ordered set. Then $x$ has an $R$-maximal element.

**3. Teichmüller-Tukey Lemma.** Let $x$ be a set, $P(X)$ a property of finite character. Let $Q = \{y : y \in (\mathbf{P}(x)) \wedge P(y)\}$. Then $Q$ has an $(\subseteq)$-maximal element.

**Theorem 2 = Exercise 114.** Hausdorff's maximal principle implies Zorn's Lemma.

Let $(x, R)$ be an inductively ordered set. According to the Hausdorff principle there exists a $(\subseteq)$-maximal subset $y$ of $x$ totally ordered by

$R$. By the inductive property $y$ has an $R$-upper bound $u$. Show that $u$ is an $R$-maximal element of $x$ (if not, show that there exists a totally ordered subset properly including $y$).

**Theorem 3 = Exercise 115.** Zorn's Lemma implies the Teichmüller-Tukey Lemma.

Show that the set $Q$ of the Teichmüller-Tukey Lemma is inductively ordered by the inclusion relation.

**Theorem 4 = Exercise 116.** The Teichmüller-Tukey Lemma implies the Hausdorff maximal principle.

Let $(x, R)$ be an ordered set; let $P(X)$ be the property

$$\text{“}R \cap (X \times X) \text{ is a relation of total order on } X\text{”}.$$

Show that $P(X)$ is of finite character and apply the Teichmüller-Tukey Lemma.

**Theorem 5 = Exercise 117.** Zorn's Lemma implies the Axiom of Choice.

Let $(x_i)_{i \in I}$ be a non-empty family of non-empty sets. Let $C$ be the class of all functional relations $f$ with $\operatorname{Dom} f \subseteq I$ such that $f(i) \in x_i$ for all indices $i$ in $I$. Show that $C$ is inductively ordered by the inclusion relation and apply Zorn's Lemma, so that $C$ has a maximal element $v$, say. Show that $\operatorname{Dom} v = I$.

**Theorem 6 = Exercise 118.** The Axiom of Choice implies the Hausdorff maximal principle.

Since the Axiom of Choice holds it follows from Zermelo's Theorem that there is a relation of well-ordering on $x$ and hence (by Exercise 96) a bijection $f$ from some ordinal number $\alpha$ onto $x$. Define a mapping $g$ from $\alpha$ to $x$ by setting for each element $\beta$ of $\alpha$

$$g(\beta) = \begin{cases} f(\beta) & \text{if for all elements } \gamma \text{ of } \beta \text{ we have} \\ & \quad \text{either } (g(\gamma), f(\beta)) \in R \text{ or } (f(\beta), g(\gamma)) \in R, \\ f(\emptyset) & \text{otherwise.} \end{cases}$$

Show that $g^{\rightarrow}(\alpha)$ is a $(\subseteq)$-maximal totally ordered subset of $x$.

The theorems in this section show that in a system in which the Axiom of Choice holds then Zermelo's Theorem (in its informal version: every set has a well- ordering), Zorn's Lemma, the Hausdorff maximal principle and the Teichmüller-Tukey Lemma all hold and that in fact these propositions are all equivalent.

**Exercise 119.** Let $(E, R)$ be an ordered set. Let

$$T = \{X : X \in \mathbf{P}(E) \text{ and } R \cap (X \times X) \text{ is a total order on } X\}.$$

Prove that every set in $T$ is included in an $(\subseteq)$-maximal set in $T$.

Let $t_0$ be any set in $T$ and show that $T_0 = \{t : (t \in T) \wedge (t \supseteq t_0)\}$ is inductively ordered.

**Exercise 120.** A **Hamel basis** is a basis for **R** (the field of real numbers) as a vector space over **Q** (the field of rational numbers), i.e. a set $H$ of real numbers such that every real number is uniquely expressible as a linear combination of finitely many elements of $H$ with rational number coefficients. Prove that the property of being a linearly independent subset of **R** over **Q** (a **Q**-free subset of **R**) is a property of finite character and deduce that there exists at least one Hamel basis.

Show that the property "$(X \subseteq R) \wedge (X$ is **Q**-free)" is of finite character and apply the Teichmüller-Tukey Lemma. (By "**Q**-free" we mean linearly independent over the field **Q** of rational numbers.)

# Chapter 11

# INFINITE SETS

*Throughout this chapter we assume that the Axiom of Choice holds.*

In Chapter 10 we defined a set to be finite if it is equipotent to a natural number; so we naturally define a set to be **infinite** if it is not finite, i.e. if it is not equipotent to any natural number.

**Theorem 1 = Exercise 121.** Every subset of a finite set is finite.

It is clearly sufficient to prove that for every natural number $n$ all subsets of $n$ are finite. To do this we apply Exercise 106 to the set $\{n : (n \in \omega) \wedge (\text{every subset of } n \text{ is finite})\}$.

**Theorem 2 = Exercise 122.** If $a$ is a finite set then $a$ is not equipotent to any proper subset of itself.

Here again it is sufficient to show that no natural number is equipotent to a proper subset of itself. To prove this we may apply Exercise 106 to the set $\{n : (n \in \omega) \wedge (n \text{ is not equipotent to any proper subset of } n)\}$.

**Exercise 123** Show that $\omega$ is infinite.

It is easy to construct a bijection from $\omega$ onto a proper subset of $\omega$.

A set is called **Dedekind-infinite** if it is equipotent to a proper subset of itself. Theorem 2 shows that Dedekind-infinite sets are indeed infinite.

**Exercise 124.** Show that a finite set $a$ is equipotent to a *unique* natural number.

Suppose that $a$ is equipotent to two distinct natural numbers $m$ and $n$. Deduce that one of them is equipotent to a proper subset of itself.

A set is said to be **denumerable** or **countably infinite** if it is equipotent to the set of natural numbers $\omega$. A set is said to be **countable** if it is either finite or denumerably infinite.

**Theorem 3 = Exercise 125.** Every infinite set $a$ includes a denumerable subset.

By Zermelo's Theorem (Exercise 111) there is a well-ordering relation $R$ on $a$. Apply Exercise 85 to the well-ordered sets $(a, R)$ and $(\omega, E \mid \omega)$.

**Theorem 4 = Exercise 126.** A set is infinite if and only if it is Dedekind-infinite.

The *if* part follows, as we remarked above, from Exercise 122. To prove the *only if* part use Exercises 125 and 123.

# Chapter 12

# CARDINALS

*Throughout this chapter we assume that the Axiom of Choice holds.*

Let $x$ be a set. Let

$$N(x) = \{\alpha : (\alpha \in On) \wedge (\alpha \text{ is equinumerous with } x)\}.$$

Since $x$ can be well-ordered (by Zermelo's Theorem) and hence is equinumerous with at least one ordinal number, $N(x)$ is a non-empty subclass of the well-ordered class $On$ and hence has an $E$-least member. This $E$-least member of $N(x)$ is called the **cardinal** of $x$ and is denoted by Card $x$.

The examples in the next exercise are immediate consequences of the definition.

**Exercise 127.** Card $\emptyset = 0$ $(= \emptyset)$; if $a$ is any set then Card $\{a\} = 1$ $(= \{\emptyset\})$; if $a$ and $b$ are distinct sets then Card $\{a, b\} = 2$ $(= \{0, 1\})$.

**Theorem 1 = Exercise 128.** Let $x$ and $y$ be sets. Then we have Card $x =$ Card $y$ if and only if $x$ is equinumerous with $y$.

**Exercise 129.** Show that a set is finite if and only if its cardinal is a natural number.

This is an immediate consequence of the definition of finiteness.

The class of cardinals is denoted by $Cn$. Since $Cn$ is a subclass of $On$ it is well-ordered (and hence totally ordered) by the restriction to $Cn$ of the order relation $E$. For cardinals $\mathfrak{a}$ and $\mathfrak{b}$ we shall write $\mathfrak{a} \leq \mathfrak{b}$ if $(\mathfrak{a}, \mathfrak{b}) \in E$ and $\mathfrak{a} < \mathfrak{b}$ if $(\mathfrak{a}, \mathfrak{b}) \in E$ and $\mathfrak{a} \neq \mathfrak{b}$.

**Exercise 130.** Let $a$ and $b$ be sets. Show that Card $a \leq$ Card $b$ if and only if there exists an injection of $a$ into $b$.

It is not hard to prove that if Card $a \leq$ Card $b$ then there is an injection from $a$ to $b$. (Distinguish the cases where Card $a =$ Card $b$ and Card $a \in$ Card $b$.)

The converse is more delicate. Let $\alpha =$ Card $a$, $\beta =$ Card $b$ and suppose there is an injection $j$ from $a$ to $b$; let $x = j^{\rightarrow}(a)$ and $\xi =$ Ord $(x, R \,|\, x)$, where $R$ is the well-ordering on $b$ induced by $E \,|\, \beta$. Show that $\xi \subseteq \beta$ and deduce that $(\xi, \beta) \in E$ (for, if not, we would have $\beta \in \xi$ and this leads quickly to a contradiction to the Axiom of Foundation).

**Theorem 2 (Schröder-Bernstein Theorem) = Exercise 131.** Let $a$ and $b$ be sets. If there exists an injection of $a$ into $b$ and an injection of $b$ into $a$ then $a$ and $b$ are equipotent.

This is an immediate consequence of Exercise 130. (This result can also be proved without using cardinals or the Axiom of Choice.)

**Theorem 3 (Cantor's Theorem) = Exercise 132.** For every set $a$ we have Card $a <$ Card $\mathbf{P}(a)$.

It is easy to show that Card $a \leq$ Card $\mathbf{P}(a)$. If Card $a =$ Card $\mathbf{P}(a)$ then there is a bijection $f$ from $a$ into $\mathbf{P}(a)$; examine the set

$$y = \{t : (t \in a) \wedge (t \notin f(t))\}.$$

**Exercise 133.** $Cn$ is a proper class.

Suppose $Cn$ were a set. Let $X = \bigcup Cn$ and show that we have Card $\mathbf{P}(X) \leq$ Card $X$ in contradiction to Cantor's Theorem.

**Exercise 134.** Let $2 = \{0, 1\} = \{\emptyset, \{\emptyset\}\}$ as in Exercise 87. Show that for every set $x$ we have Card $\mathbf{P}(x) =$ Card (Map $(x, 2)$).

Make use of the characteristic functions $\chi_a$ of the subsets $a$ of $x$ defined by setting

$$\chi_a(t) = \begin{cases} 1 & \text{if } t \in a \\ 0 & \text{if } t \notin a. \end{cases}$$

Let $S$ be the relation on $On \sim \omega$ defined by setting $(\alpha, \beta) \in S$ if and only if $\alpha$ is equinumerous with $\beta$. Then $S$ is an equivalence relation. It can be shown that for each infinite ordinal $\alpha$ the $S$-equivalence class $S(\alpha)$ is a set. The $E$-least member of $S(\alpha)$ is called an **initial ordinal.** Thus an ordinal number $\beta$ is an initial ordinal if and only if for every ordinal number $\gamma$ such that $\gamma \in \beta$ there exists an injection but not a bijection from $\gamma$ to $\beta$.

The class of initial ordinals in denoted by $Io$; clearly $Cn = Io \cup \omega$. It follows that $Io$ is a proper class. It can be shown that there is an isomorphism $F$ from $On$ onto $Io$; for each ordinal number $\alpha$ we write $F(\alpha) = \omega_\alpha =$ the $\alpha$-th initial ordinal. The least initial ordinal is thus $\omega_0 = \omega$; the next, which is the least uncountable ordinal is $\omega_1$. We write $\Omega = \omega_1^+ = \omega_1 \cup \{\omega_1\}$; the ordinal $\Omega$ is used to construct interesting topological spaces such as the ordinal spaces $[0, \Omega)$ and $[0, \Omega]$, the long line and the Tihonov plank and corkscrew.

# Chapter 13

# CARDINAL AND ORDINAL ARITHMETIC

Let $(\mathfrak{a}_i)_{i \in I}$ be a family of cardinals indexed by a set $I$. The **cardinal sum** or simply the **sum** $\sum_{i \in I} \mathfrak{a}_i$ of $(\mathfrak{a}_i)_{i \in I}$ is defined to be the cardinal of the union $\bigcup_{i \in I}(\mathfrak{a}_i \times \{i\})$. If $\mathfrak{a}$ and $\mathfrak{b}$ are cardinals then the sum $\mathfrak{a}+\mathfrak{b}$ of $\mathfrak{a}$ and $\mathfrak{b}$ is defined to be the cardinal of $(\mathfrak{a} \times \{0\}) \cup (\mathfrak{b} \times \{1\})$. Notice that in these definitions the sets forming the unions are equipotent to the cardinals being added but (thanks to their second factors) are pairwise disjoint.

The **cardinal product** $\mathbf{P}_{i \in I}\mathfrak{a}_i$ of the family of cardinals $(\mathfrak{a}_i)_{i \in I}$ is defined by setting

$$\mathbf{P}_{i \in I}\mathfrak{a}_i = \text{Card} \prod_{i \in I} \mathfrak{a}_i.$$

Similarly if $\mathfrak{a}$ and $\mathfrak{b}$ are cardinals then we define the product $\mathfrak{a} \cdot \mathfrak{b}$ of $\mathfrak{a}$ and $\mathfrak{b}$ to be Card $(\mathfrak{a} \times \mathfrak{b})$.

**Exercise 135.** Let $\mathfrak{a}$ and $\mathfrak{b}$ be cardinals, $A$ and $B$ disjoint sets equipotent to $\mathfrak{a}$ and $\mathfrak{b}$ respectively. Show that $\mathfrak{a} + \mathfrak{b} = \text{Card}\ (A \cup B)$.

**Exercise 136.** Let $\mathfrak{a}$ and $\mathfrak{b}$ be cardinals, $A$ and $B$ sets equipotent to $\mathfrak{a}$ and $\mathfrak{b}$ respectively. Show that $\mathfrak{a} \cdot \mathfrak{b} = \text{Card}\ (A \times B)$.

For Exercises 135 and 136 notice that if we have bijections from $\mathfrak{a}$ to $A$ and from $\mathfrak{b}$ to $B$ then it is easy to construct bijections from $(\mathfrak{a} \times \{0\}) \cup (\mathfrak{b} \times \{1\})$ to $A \cup B$ and from $\mathfrak{a} \times \mathfrak{b}$ to $A \times B$.

The results of Exercise 137 follow either immediately from the definitions of sum and product or by appropriate applications of Exercises 135 and 136.

**Exercise 137.** Show that for all cardinals $\mathfrak{a}$, $\mathfrak{b}$, $\mathfrak{c}$ we have
(1) $\mathfrak{a} + 0 = \mathfrak{a} \cdot 1 = \mathfrak{a}$;
(2) $\mathfrak{a} \cdot 0 = 0$;
(3) $\mathfrak{a} + \mathfrak{b} = \mathfrak{b} + \mathfrak{a}$ and $\mathfrak{a} \cdot \mathfrak{b} = \mathfrak{b} \cdot \mathfrak{a}$;
(4) $\mathfrak{a} + (\mathfrak{b} + \mathfrak{c}) = (\mathfrak{a} + \mathfrak{b}) + \mathfrak{c}$ and $\mathfrak{a} \cdot (\mathfrak{b} \cdot \mathfrak{c}) = (\mathfrak{a} \cdot \mathfrak{b}) \cdot \mathfrak{c}$;
(5) $\mathfrak{a} \cdot (\mathfrak{b} + \mathfrak{c}) = \mathfrak{a} \cdot \mathfrak{b} + \mathfrak{a} \cdot \mathfrak{c}$.

**Exercise 138.** If $\mathfrak{a}$ and $\mathfrak{b}$ are cardinals such that $\mathfrak{a} + 1 = \mathfrak{b} + 1$ show that $\mathfrak{a} = \mathfrak{b}$.

Express $\mathfrak{a} + 1 = \mathfrak{b} + 1$ in two ways as the union of a subset and a singleton.

**Exercise 139.** Prove that for every natural number $n$ we have $n + 1 = n^+$.

This follows at once from the definitions of addition and successor.

Let $\mathfrak{a}$ and $\mathfrak{b}$ be cardinals. Then the $\mathfrak{b}$-th power of $\mathfrak{a}$, denoted by $\mathfrak{a}^{\mathfrak{b}}$, is defined to be the cardinal of the set $\mathrm{Map}(\mathfrak{b}, \mathfrak{a})$.

**Exercise 140.** Let $\mathfrak{a}$ and $\mathfrak{b}$ be cardinals. If $A$ and $B$ are sets equipotent to $\mathfrak{a}$ and $\mathfrak{b}$ respectively prove that $\mathfrak{a}^{\mathfrak{b}} = \mathrm{Card}\ \mathrm{Map}(B, A)$.

Using given bijections from $\mathfrak{a}$ to $A$ and $\mathfrak{b}$ to $B$ it is easy to define a bijection from $\mathrm{Map}\,(\mathfrak{b}, \mathfrak{a})$ to $\mathrm{Map}\,(B, A)$.

**Exercise 141.** (1) Let $\mathfrak{a}$ be any cardinal. Prove that $\mathfrak{a}^0 = 1$, $\mathfrak{a}^1 = \mathfrak{a}$, $1^{\mathfrak{a}} = 1$, $2 \cdot \mathfrak{a} = \mathfrak{a} + \mathfrak{a}$, $\mathfrak{a}^2 = \mathfrak{a} \cdot \mathfrak{a}$ and that $0^{\mathfrak{a}} = 0$ if $\mathfrak{a} \neq 0$.

(2) Show that if $\mathfrak{a}$, $\mathfrak{b}$, $\mathfrak{c}$ are cardinals such that $\mathfrak{a} \leq \mathfrak{b}$ then $\mathfrak{a}+\mathfrak{c} \leq \mathfrak{b}+\mathfrak{c}$ and $\mathfrak{a} \cdot \mathfrak{c} \leq \mathfrak{b} \cdot \mathfrak{c}$.

The results of Exercise 141 follow easily from the definitions of the terms involved; for the last part of (1) we refer to Exercise 27. For (2) we use Exercise 130.

**Exercise 142.** Let $\mathfrak{a}$, $\mathfrak{b}$, $\mathfrak{c}$ be cardinals. Show that
(1) $\mathfrak{a}^{\mathfrak{b}+\mathfrak{c}} = \mathfrak{a}^{\mathfrak{b}} \cdot \mathfrak{a}^{\mathfrak{c}}$;
(2) $\mathfrak{a}^{\mathfrak{b}\cdot\mathfrak{c}} = (\mathfrak{a}^{\mathfrak{b}})^{\mathfrak{c}}$;
(3) $(\mathfrak{a} \cdot \mathfrak{b})^{\mathfrak{c}} = \mathfrak{a}^{\mathfrak{c}} \cdot \mathfrak{b}^{\mathfrak{c}}$.

For Part (1) let $A$ be equipotent to $\mathfrak{a}$, $B$ and $C$ disjoint sets equipotent to $\mathfrak{b}$ and $\mathfrak{c}$ respectively and define a bijection from Map $(B \cup C, A)$ to Map $(B, A) \times$ Map $(C, A)$. For Parts (2) and (3) we use Exercises 38 and 39 respectively.

**Exercise 143.** Let $\mathfrak{a}$ be a cardinal, $(\mathfrak{b}_i)_{i\in I}$ a family of cardinals indexed by a set $I$. Prove that $\mathfrak{a}^{\sum_{i\in I} \mathfrak{b}_i} = \mathbf{P}_{i\in I}\mathfrak{a}^{\mathfrak{b}_i}$.

This is a generalisation to arbitrary products of Part (1) of Exercise 142.

**Exercise 144.** If $A$ is any set and $\mathfrak{a} =$ Card $A$ show that the cardinal of $\mathbf{P}(A)$ is $2^{\mathfrak{a}}$.

Use Exercise 131.

It can be shown, though we shall not do so here, that the set $\mathbf{R}$ of real numbers (which we do not define formally in this book) is equipotent to $\mathbf{P}(\omega)$. Hence $\operatorname{Card}\mathbf{R} = \operatorname{Card}\mathbf{P}(\omega) > \operatorname{Card}\omega$ (by Cantor's Theorem, Exercise 132) and so the set of real numbers is not denumerable.

**Exercise 145.** Show that Card $\omega = \omega$.

When we think of $\omega$ as a cardinal rather than as an ordinal we denote it by $\aleph_0$ (read "aleph-nought").

**Exercise 146.** Prove that for every natural number $n$ we have $\aleph_0 + n = \aleph_0$ and that for all non-zero natural numbers $n$ we have $n \cdot \aleph_0 = \aleph_0$.

**Exercise 147.** Let $\mathfrak{a}$ be an infinite cardinal such that $\mathfrak{a}^2 = \mathfrak{a}$. Prove that $\mathfrak{a} = 2 \cdot \mathfrak{a} = 3 \cdot \mathfrak{a}$.

**Exercise 148.** Prove that $\aleph_0 + \aleph_0 = \aleph_0 \cdot \aleph_0 = \aleph_0$.

It is easy to set up a bijection from $(\aleph_0 \times \{0\}) \cup (\aleph_0 \times \{1\})$ to $\aleph_0$.

The informal proof that $\aleph_0 \cdot \aleph_0 = \aleph_0$ (much more convincing than its formalisation) is exhibited in the diagram

$$\begin{array}{ccccc} 0 & 2 & 5 & 9 & 14 \\ 1 & 4 & 8 & 13 & \dots \\ 3 & 7 & 12 & \dots & \dots \\ 6 & 11 & \dots & \dots & \dots \\ 10 & \dots & \dots & \dots & \dots \end{array}$$

To make this formal we must define a bijection $f$ from $\aleph_0 \times \aleph_0$ to $\aleph_0$ such that for all ordered pairs $(i, j)$ of natural numbers $f(i, j)$ is the natural number in the $(i, j)$-th position.

**Exercise 149.** Prove that for every infinite cardinal $\mathfrak{a}$ we have $\mathfrak{a}^2 = \mathfrak{a} \cdot \mathfrak{a} = \mathfrak{a}$.

Consider the set of ordered pairs $(A, f)$ where $A \subseteq \mathfrak{a}$ and $f$ is a bijection from $A$ to $A \times A$. Show that this set is inductively ordered by $R$ where $((A, f), (B, g)) \in R$ if and only if $A \subseteq B$ and $g \mid A = f$. By Zorn's Lemma there is an $R$-maximal element $(M, k)$. Show that $\operatorname{Card} M = \mathfrak{a}$.

Let $(X_1, R_1)$, $(X_2, R_2)$ be two well-ordered sets. Let $X = X_1 \oplus X_2 = (X_1 \times \{1\}) \cup (X_2 \times \{2\})$. Define a relation $R = R_1 \oplus R_2$ on $X$ by setting

$$((x, i), (y, j)) \in R \iff (i < j) \vee ((i = j) \wedge ((x, y) \in R_i)).$$

**Exercise 150.** Prove that $R$ is a well-ordering on $X$.

Let $\alpha$ and $\beta$ be ordinal numbers. Then the **ordinal sum** of $\alpha$ and $\beta$ is defined to be the ordinal number of $(\alpha \oplus \beta, E \mid \alpha \oplus E \mid \beta)$.

Again let $(X_1, R_1)$, $(X_2, R_2)$ be well-ordered sets. Define a relation $S = R_1 \otimes R_2$ on $X_1 \times X_2$ by setting

$$\begin{aligned} &((x_1, x_2), (y_1, y_2)) \in S \iff \\ &\qquad ((x_2 \neq y_2) \wedge ((x_2, y_2) \in R_2)) \vee ((x_2 = y_2) \wedge ((x_1, y_1) \in R_1)). \end{aligned}$$

**Exercise 151.** Prove that $S$ is a well-ordering on $X_1 \times X_2$.

$S$ is called the **reverse lexical order** on $X_1 \times X_2$.

Let $\alpha$ and $\beta$ be ordinal numbers. Then the **ordinal product** $\alpha \cdot \beta$ of $\alpha$ and $\beta$ is defined to be the ordinal number of $(\alpha \times \beta, (E \mid \alpha) \otimes (E \mid \beta))$.

**Exercise 152.** Show that if $(X_1, R_1)$ is isomorphic to $(X_1', R_1')$ and $(X_2, R_2)$ is isomorphic to $(X_2', R_2')$ then $(X_1 \oplus X_2, R_1 \oplus R_2)$ is isomorphic to $(X_1' \oplus X_2', R_1' \oplus R_2')$ and $(X_1 \times X_2, R_1 \otimes R_2)$ is isomorphic to $(X_1' \times X_2', R_1' \otimes R_2')$.

**Exercise 153.** Show that for all ordinal numbers $\alpha$, $\beta$, $\gamma$ we have
(1) $\alpha + 0 = \alpha = 0 + \alpha = \alpha \cdot 1 = 1 \cdot \alpha$;
(2) $0 \cdot \alpha = \alpha \cdot 0 = 0$;
(3) $\alpha + (\beta + \gamma) = (\alpha + \beta) + \gamma$ and $\alpha \cdot (\beta \cdot \gamma) = (\alpha \cdot \beta) \cdot \gamma$;
(4) $\alpha \cdot (\beta + \gamma) = \alpha \cdot \beta + \alpha \cdot \gamma$.

**Exercise 154.** Show that $1 + \omega = \omega$ but that $\omega + 1 \neq \omega$.

**Exercise 155.** Show that $2 \cdot \omega = \omega$ but that $\omega \cdot 2 = \omega + \omega \neq \omega$.

To prove the results of the last three exercises we use Exercise 152 and Exercise 97.

# Part II

# ANSWERS

# Chapter 14

# ANSWERS TO CHAPTER 1

1. Let $C$ be a class. Then we have

$$\begin{aligned} C \in \bigcap \emptyset &\iff C \text{ is a set and } C \text{ belongs to every member of } \emptyset \\ &\iff C \text{ is a set} \\ &\iff C \in V. \end{aligned}$$

   Thus $\bigcap \emptyset = V$ by the Axiom of Extensionality.

   Again let $C$ be a class. Then

$$\begin{aligned} C \in \bigcup \emptyset &\iff C \text{ is a set and there is} \\ &\qquad\qquad \text{a member } x \text{ of } \emptyset \text{ such that } C \in x \\ &\iff C \in \emptyset \end{aligned}$$

   (since $\emptyset$ has no members).

   So $\bigcup \emptyset = \emptyset$ by the Axiom of Extensionality.

   (Perhaps the second part is more convincingly shown by contradiction: suppose $\bigcup \emptyset \neq \emptyset$; then there is a set $C$ which belongs to some member of $\emptyset$: this is a contradiction since $\emptyset$ has no members.)

2. Since $A \subseteq B$ and $B \subseteq C$ we have $A \subseteq C$. Since $A \subseteq C$ and $C \subseteq A$ we have $A = C$.

   Similarly $B = C$.

3. Since $A \subset B$ and $B \subset C$ we certainly have $A \subseteq B$ and $B \subseteq C$, and so $A \subseteq C$.

Since $B \subset C$ there is an element $c$ such that $c \in C$ and $c \notin B$. Then, since $A \subset B$, we have that $c \notin A$. (Actually we are using only the weaker hypothesis that $A \subseteq B$.) So $A \subset C$.

4. (a) Suppose $A \subseteq B$. Let $x$ be any set.

   Then $x \in (A \cup B) \Longrightarrow ((x \in A) \ \vee \ (x \in B)) \Longrightarrow x \in B$; so $A \cup B \subseteq B$.

   But $x \in B \Longrightarrow ((x \in A) \ \vee \ (x \in B)) \Longrightarrow x \in A \cup B$; so $B \subseteq A \cup B$. Hence $A \cup B = B$.

   Conversely, suppose $A \cup B = B$. Let $x$ be any set.

   Then $x \in A \Longrightarrow ((x \in A) \vee (x \in B)) \Longrightarrow x \in A \cup B (= B)$. Hence $A \subseteq B$.

   (b) Suppose $A \subseteq B$. Let $x$ be any set.

   Then $x \in A \cap B \Longrightarrow ((x \in A) \ \wedge \ (x \in B)) \Longrightarrow x \in A$; so $A \cap B \subseteq A$. But $x \in A \Longrightarrow ((x \in A) \ \wedge \ (x \in B)) \Longrightarrow x \in A \cap B$; so $A \subseteq A \cap B$.

   Hence $A \cap B = A$.

   Conversely, suppose $A \cap B = A$. Let $x$ be any set. Then

   $x \in A \Longrightarrow x \in A \cap B \Longrightarrow ((x \in A) \ \wedge \ (x \in B)) \Longrightarrow x \in B$.

   Hence $A \subseteq B$.

   (c) Suppose $A \subseteq B$. Let $x$ be any set.

   Then $x \in \sim B \Longrightarrow x \notin B \Longrightarrow x \notin A \Longrightarrow x \in \sim A$. So $\sim B \subseteq \sim A$.

   Conversely, suppose $\sim B \subseteq \sim A$. Let $x$ be any set.

   Then $x \in A \Longrightarrow x \notin \sim A \Longrightarrow x \notin \sim B \Longrightarrow x \in B$.

   Hence $A \subseteq B$.

   (d) Suppose $A \subseteq B$.

   If $A \cap (\sim B) \neq \emptyset$ then there is a set $x$ such that $x \in A$ and $x \in \sim B$, i.e. $x \in A$ and $x \notin B$. This is a contradiction.

   So $A \cap (\sim B) = \emptyset$.

   Conversely, suppose $A \cap (\sim B) = \emptyset$.

   If it is false that $A \subseteq B$ then there is an element $a$ of $A$ which does not belong to $B$ and hence belongs to $\sim B$. So $A \cap (\sim B) \neq \emptyset$. This is a contradiction. So $A \subseteq B$.

5. Suppose $A \subseteq B$.

   Let $x$ be any element of $\bigcup A$. Then there is an element $a$ of $A$ such that $x \in a$. Since $A \subseteq B$, we have $a \in B$. Hence $x \in \bigcup B$. Thus $\bigcup A \subseteq \bigcup B$.

   Let $x$ be any element of $\bigcap B$. Then $x$ belongs to every element of $B$ and hence in particular to every element of $A$. Thus $x \in \bigcap A$. So $\bigcap B \subseteq \bigcap A$.

6. Suppose $a \in B$.

   Let $x$ be any element of $a$. Since $x \in a$ and $a \in B$ it follows that $x \in \bigcup B$. So $a \subseteq \bigcup B$.

   Let $x$ be any element of $\bigcap B$. Then $x$ belongs to every element of $B$; in particular, $x \in a$. So $\bigcap B \subseteq a$.

7. (a) Suppose $\bigcap V \neq \emptyset$; let $x$ be an element of $\bigcap V$. Then $x$ is a set. Now $\emptyset \subseteq x$, so it follows that $\emptyset$ is a set, i.e. $\emptyset \in V$. Thus $x$, which belongs to every element of $V$, belongs to $\emptyset$. This is a contradiction; so $\bigcap V = \emptyset$.

   (b) Let $x \in \bigcup V$. Then $x$ is a set and so $x \in V$. Thus $\bigcup V \subseteq V$. Conversely, suppose $x \in V$. Then $x$ is a set; so is $\mathbf{P}(x)$ and since $x \subseteq x$ we have $x \in \mathbf{P}(x)$. Thus $x$ is in one of the elements of $V$ and so $x \in \bigcup V$. Thus $V \subseteq \bigcup V$.

   (c) Let $x \in \mathbf{P}(V)$. Then $x$ is a set and so $x \in V$. So $\mathbf{P}(V) \subseteq V$. Conversely, let $x \in V$. Then $x$ is a set and $x \subseteq V$. So $x \in \mathbf{P}(V)$. Thus $V \subseteq \mathbf{P}(V)$.

8. Suppose $A \neq \emptyset$. Then $A$ has an element, $a$ say, which is a set.

   Now $\bigcap A \subseteq a$ (see Exercise 6) and hence $\bigcap A$, being a subclass of a set, is a set.

9. Let $t$ be any set. Then $t \in x \cup y \iff (t \in x) \vee (t \in y) \iff t \in$ one of the elements of $\{x, y\} \iff t \in \bigcup\{x, y\}$.

   Let $t$ be any set. Then $t \in \{x\} \cup \{y\} \iff (t = x) \vee (t = y) \iff t \in \{x, y\}$

# Chapter 15

# ANSWERS TO CHAPTER 2

10. If $a = c$ and $b = d$ then certainly $(a, b) = (c, d)$.

    On the other hand, suppose we have $(a, b) = (c, d)$, i.e. $\{\{a\}, \{a, b\}\} = \{\{c\}, \{c, d\}\}$.

    Case 1. $a = b$. In this case we have $\{\{a\}, \{a, b\}\} = \{\{a\}, \{a, a\}\} = \{\{a\}\}$. Hence $\{c\} = \{a\}$ and $\{c, d\} = \{a\}$; it follows that $c = a$ and $d = a = b$.

    Case 2. $a \neq b$. In this case we have $\{a, b\} \neq \{a\}$ and so $\{c\} \neq \{c, d\}$, whence $d \neq c$. Now either $\{a\} = \{c\}$, whence $a = c$ or $\{a\} = \{c, d\}$, so that $a = c = d$, which is a contradiction. Hence $\{a\} = \{c\}$, so that $a = c$, and thus $\{a, b\} = \{c, d\}$, from which we deduce eventually that $b = d$.

11. If $A \times B \neq \emptyset$ then $A \times B$ has an element, $w$ say; since $w \in A \times B$ there are elements $a$ of $A$ and $b$ of $B$ such that $w = (a, b)$. Thus $A \neq \emptyset$ and $B \neq \emptyset$.

    Conversely, if $A \neq \emptyset$ and $B \neq \emptyset$ there are elements $a$ of $A$ and $b$ of $B$. Then $(a, b) \in A \times B$, which is therefore non-empty.

12. Since $A$ and $B$ are non-empty it follows that $A \times B$ is non-empty.

    Suppose $A \subseteq A'$, $B \subseteq B'$. Let $(a, b)$ be any element of $A \times B$. Then $a \in A$ and $b \in B$; so $a \in A'$ and $b \in B'$; hence we have $(a, b) \in A' \times B'$. Thus $A \times B \subseteq A' \times B'$.

    Conversely, suppose that $A \times B \subseteq A' \times B'$. Let $a$ be any element of $A$, $b$ any element of $B$. Then $(a, b) \in A \times B$ and so $(a, b) \in A' \times B'$. Thus $a \in A'$ and $b \in B'$. Hence $A \subseteq A'$ and $B \subseteq B'$.

13. (1) We have $(a, x) \in A \times (B \cup C)$

$$\begin{aligned}
&\iff (a \in A) \wedge (x \in B \cup C) \\
&\iff (a \in A) \wedge ((x \in B) \vee (x \in C)) \\
&\iff ((a \in A) \wedge (x \in B)) \vee ((a \in A) \wedge (x \in C)) \\
&\iff ((a, x) \in A \times B) \vee ((a, x) \in A \times C) \\
&\iff (a, x) \in (A \times B) \cup (A \times C).
\end{aligned}$$

(2) We have $(a, x) \in A \times (B \cap C)$

$$\begin{aligned}
&\iff (a \in A) \wedge (x \in B \cap C) \\
&\iff (a \in A) \wedge ((x \in B) \wedge (x \in C)) \\
&\iff ((a \in A) \wedge (x \in B)) \wedge ((a \in A) \wedge (x \in C)) \\
&\iff ((a, x) \in A \times B) \wedge ((a, x) \in A \times C) \\
&\iff (a, x) \in (A \times B) \cap (A \times C).
\end{aligned}$$

(3) We have $(a, x) \in A \times (B \sim C)$

$$\begin{aligned}
&\iff (a \in A) \wedge (x \in B \sim C) \\
&\iff (a \in A) \wedge (x \in B) \wedge (x \notin C) \\
&\iff ((a \in A) \wedge (x \in B)) \wedge ((a \in A) \wedge (x \notin C)) \\
&\iff ((a, x) \in A \times B) \wedge ((a, x) \notin A \times C) \\
&\iff (a, x) \in (A \times B) \sim (A \times C).
\end{aligned}$$

14. $\bigcap(a, b) = \bigcap\{\{a\}, \{a, b\}\} = \{a\}$; $\bigcap\bigcap(a, b) = a$; $\bigcup\bigcap(a, b) = a$.

$\bigcup(a, b) = \bigcup\{\{a\}, \{a, b\}\} = \{a, b\}$.

$\bigcup\bigcup(a, b) = a \cup b$; $\bigcap\bigcup(a, b) = a \cap b$.

$\left(\bigcap\bigcup(a, b)\right) \cup \left(\bigcup\bigcup(a, b) \sim \bigcup\bigcap(a, b)\right) = (a \cap b) \cup ((a \cup b) \sim a) = b$.

15. Let $(x, y) \in R$. Recall that $(x, y) = \{\{x\}, \{x, y\}\}$. Then $x \in \bigcup\bigcup R$ and $y \in \bigcup\bigcup R$. It follows that $\text{Dom}R \subseteq \bigcup\bigcup R$ and Range $R \subseteq \bigcup\bigcup R$.

Since $R$ is a set we know that $\bigcup R$ and $\bigcup\bigcup R$ are sets (by the Union Axiom) and hence Dom $R$ and Range $R$ are sets.

16. We have $(z, x) \in (S \circ R)^{-1}$

$$\begin{aligned}
&\iff (x, z) \in S \circ R \\
&\iff (\exists y)(((x, y) \in R) \wedge ((y, z) \in S)) \\
&\iff (\exists y)(((z, y) \in S^{-1}) \wedge ((y, x) \in R^{-1})) \\
&\iff (z, x) \in R^{-1} \circ S^{-1}.
\end{aligned}$$

So $(S \circ R)^{-1} = R^{-1} \circ S^{-1}$.

17. $(x, t) \in T \circ (S \circ R)$
$$\iff (\exists z)(((x, z) \in S \circ R) \wedge ((z, t) \in T))$$
$$\iff (\exists z)(((\exists y)((x, y) \in R) \wedge ((y, z) \in S)) \wedge ((z, t) \in T))$$
$$\iff (\exists y)(((x, y) \in R) \wedge ((\exists z)(((y, z) \in S) \wedge ((z, t) \in T))))$$
$$\iff (\exists y)(((x, y) \in R) \wedge ((y, z) \in T \circ S))$$
$$\iff (x, t) \in (T \circ S) \circ R.$$

Thus $T \circ (S \circ R) = (T \circ S) \circ R$.

18. We have $(x, y) \in (R \cap S)^{-1}$
$$\iff (y, x) \in R \cap S$$
$$\iff ((y, x) \in R) \wedge ((y, x) \in S)$$
$$\iff ((x, y) \in R^{-1}) \wedge ((x, y) \in S^{-1})$$
$$\iff (x, y) \in R^{-1} \cap S^{-1}.$$

So $(R \cap S)^{-1} = R^{-1} \cap S^{-1}$. Similarly $(R \cup S)^{-1} = R^{-1} \cup S^{-1}$.

19. Suppose $A \subseteq$ Dom $R$.

Let $a$ be any element of $A$. Then there is an element $b$ such that $(a, b) \in R$. Since $a \in A$ and $(a, b) \in R$ we see that $b \in R^{\rightarrow}(A)$. Since $b \in R^{\rightarrow}(A)$ and $(a, b) \in R$ we have $a \in R^{\leftarrow}(R^{\rightarrow}(A))$. So $A \subseteq R^{\leftarrow}(R^{\rightarrow}(A))$.

Conversely, suppose that $A \subseteq R^{\leftarrow}(R^{\rightarrow}(A))$.

Let $a$ be any element of $A$. Then $a \in R^{\leftarrow}(R^{\rightarrow}(A))$ and so there is an element $b$ such that $(a, b) \in R$. Thus $a \in$ Dom $R$. So $A \subseteq$ Dom $R$.

20. Suppose Dom $S \subseteq$ Dom $R$.

Let $(a, b) \in S$. Then $a \in$ Dom $S$ and so $a \in$ Dom $R$; thus there is an element $t$ such that $(a, t) \in R$ and so $(t, a) \in R^{-1}$. Then, since $(a, t) \in R$, $(t, a) \in R^{-1}$, $(a, b) \in S$ we have $(a, b) \in S \circ R^{-1} \circ R$.

So $S \subseteq S \circ R^{-1} \circ R$.

Conversely, suppose that $S \subseteq S \circ R^{-1} \circ R$. Let $a \in$ Dom $S$. Then there is an element $b$ such that we have $(a, b) \in S$ and hence $(a, b) \in S \circ R^{-1} \circ R$. So there are elements $x$, $y$ such that $(a, x) \in R$, $(x, y) \in R^{-1}$ and $(y, b) \in S$. In particular, since $(a, x) \in R$ we have $a \in$ Dom $R$.

So Dom $S \subseteq$ Dom $R$.

21. (1) $(a,c) \in R \circ (A \times B)$
$$\iff (\exists b)((b \in B) \wedge ((a,b) \in A \times B) \wedge ((b,c) \in R))$$
$$\iff (\exists b)((b \in B) \wedge ((b,c) \in R) \wedge (a \in A))$$
$$\iff (a \in A) \wedge (c \in R^{\rightarrow}(B))$$
$$\iff (a,c) \in A \times R^{\rightarrow}(B).$$

(2) $(b,d) \in (C \times D) \circ R$
$$\iff (\exists c)((c \in C) \wedge ((b,c) \in R) \wedge ((c,d) \in C \times D))$$
$$\iff (\exists c)((c \in C) \wedge ((b,c) \in R)) \wedge (d \in D)$$
$$\iff (b \in R^{\leftarrow}(C)) \wedge (d \in D)$$
$$\iff (b,d) \in R^{\leftarrow}(C) \times D.$$

22. (1) $(x,y) \in (R')^{-1}$
$$\iff (y,x) \in R'$$
$$\iff ((y,x) \in \text{Dom } R \times \text{Range } R) \wedge ((y,x) \notin R)$$
$$\iff ((y,x) \in \text{Range } R^{-1} \times \text{Dom } R^{-1}) \wedge ((y,x) \notin R)$$
$$\iff ((x,y) \in \text{Dom } R^{-1} \times \text{Range } R^{-1}) \wedge ((x,y) \notin R^{-1})$$
$$\iff (x,y) \in (R^{-1})'.$$

(2) Suppose $A \supseteq \text{Dom } R$ and $B \supseteq \text{Range } R$.
Then $(x,y) \in R \circ (R^{-1})'$
$$\iff (x,y) \in R \circ (R')^{-1}$$
$$\iff (\exists t)(((x,t) \in (R')^{-1}) \wedge ((t,y) \in R))$$
$$\iff (\exists t)(((t,x) \in R') \wedge ((t,y) \in R))$$
$$\implies (x \in \text{Range } R) \wedge (y \in \text{Range } R) \wedge (x \neq y)$$
$$\implies (x,y) \in (B \times B) \sim D_B = (D_B)'.$$
Thus $R \circ (R^{-1})' \subseteq (D_B)'$.

(3) Similarly $(R^{-1})' \circ R \subseteq (D_A)'$.

# Chapter 16

# ANSWERS TO CHAPTER 3

23. Let $R$ be functional and let $Y$ be any class.

    Let $b$ be any element of $R^{\rightarrow}(R^{\leftarrow}(Y))$. Then there is an element $a$ of $R^{\leftarrow}(Y)$ such that $(a, b) \in R$. Since $a \in R^{\leftarrow}(Y)$ there is an element $y$ of $Y$ such that $(a, y) \in R$. Since $R$ is functional and both $(a, b)$ and $(a, y)$ belong to $R$ it follows that $b = y$. So $b \in Y$.

    Thus $R^{\rightarrow}(R^{\leftarrow}(Y)) \subseteq Y$.

    Conversely, suppose that for every class $Y$ the inclusion $R^{\rightarrow}(R^{\leftarrow}(Y)) \subseteq Y$ holds.

    Let $a$ be any element of Dom $R$ and suppose that $(a, b_1) \in R$ and $(a, b_2) \in R$. Let $Y = \{b_1\}$. Then $a \in R^{\leftarrow}(Y)$ and hence $b_1$ and $b_2$ both belong to $R^{\rightarrow}(R^{\leftarrow}(Y))$. Hence $\{b_1, b_2\} \subseteq R^{\rightarrow}(R^{\leftarrow}(Y)) \subseteq Y = \{b_1\}$ and so $b_1 = b_2$.

    Thus $R$ is functional.

24. (a) Suppose $R$ is functional. Let $X$ and $Y$ be any classes.

    Let $a$ be any element of $R^{\leftarrow}(X \cap Y)$. Then there is an element $b$ of $X \cap Y$ such that $(a, b) \in R$. Since $(a, b) \in R$ and $b \in X$ we have $a \in R^{\leftarrow}(X)$; since $(a, b) \in R$ and $b \in Y$ we have $a \in R^{\leftarrow}(Y)$. Hence $a \in R^{\leftarrow}(X) \cap R^{\leftarrow}(Y)$.

    Thus $R^{\leftarrow}(X \cap Y) \subseteq R^{\leftarrow}(X) \cap R^{\leftarrow}(Y)$.

    Conversely, let $a$ be any element of $R^{\leftarrow}(X) \cap R^{\leftarrow}(Y)$. Then there are elements $b_X$ of $X$ and $b_Y$ of $Y$ such that $(a, b_X) \in R$ and

$(a, b_Y) \in R$. Since $R$ is functional we have $b_X = b_Y$, which is thus an element of $X \cap Y$. Hence $a \in R^{\leftarrow}(X \cap Y)$.

Thus $R^{\leftarrow}(X) \cap R^{\leftarrow}(Y) \subseteq R^{\leftarrow}(X \cap Y)$.

(b) Suppose that for all classes $X$, $Y$ we have $R^{\leftarrow}(X \cap Y) = R^{\leftarrow}(X) \cap R^{\leftarrow}(Y)$.

Let $X$ and $Y$ be classes such that $X \cap Y = \emptyset$. Then it follows that $R^{\leftarrow}(X) \cap R^{\leftarrow}(Y) = R^{\leftarrow}(X \cap Y) = R^{\leftarrow}(\emptyset) = \emptyset$.

(c) Suppose that for all classes $X$ and $Y$ for which $X \cap Y = \emptyset$ we have $R^{\leftarrow}(X) \cap R^{\leftarrow}(Y) = \emptyset$.

Let $a$ be any element of Dom $R$.

Suppose there were elements $b_1$, $b_2$ such that $b_1 \neq b_2$, $(a, b_1) \in R$ and $(a, b_2) \in R$. Let $X = \{b_1\}$, $Y = \{b_2\}$. Then $X \cap Y = \emptyset$. But $a \in R^{\leftarrow}(X) \cap R^{\leftarrow}(Y)$, which is therefore non-empty. This is a contradiction.

Hence $R$ is functional.

25. Suppose $f = f'$.

Then $(A, B) = (A', B')$ and $R = R'$. Hence $A = A'$, $B = B'$ and $R = R'$. Let $a$ be any element of $A$. Then $(a, f(a)) \in R$; so $(a, f(a)) \in R'$. But $f'(a)$ is the unique element $b'$ of $B'$ such that $(a, b') \in R'$. So $f(a) = f'(a)$.

Conversely, suppose that $A = A'$, $B = B'$ and $f(a) = f'(a)$ for all elements $a$ of $A$.

We have to show that $R = R'$. So let $(a, b)$ be any element of $R$. Then $b = f(a) = f'(a)$. Thus $(a, b) \in R'$. Hence $R \subseteq R'$ and similarly $R' \subseteq R$ and so $R = R'$. It follows at once that $f = f'$.

26. Let $f$ be a mapping from $A$ to $B$. Then $f \in \{(A, B)\} \times \mathbf{P}(A \times B)$. Since $A$ and $B$ are sets, so are $(A, B)$ (by the Pairing Axiom), $\{(A, B)\}$ (by the Pairing Axiom again), $A \times B$ (by the argument preceding Exercise 10) and $\mathbf{P}(A \times B)$ (by the Power Set Axiom). Hence $\{(A, B)\} \times \mathbf{P}(A \times B)$ is a set.

Thus since Map $(A, B) \subseteq \{(A, B)\} \times \mathbf{P}(A \times B)$ we have that Map $(A, B)$ is a set, by the Power Set Axiom.

27. (a) Suppose $A \neq \emptyset$ and $B = \emptyset$.

    Then $A \times B = \emptyset$ and so the only relation between $A$ and $B$ is the empty relation, which does not have domain $A$. So $((A, B), \emptyset)$ is not a mapping from $A$ to $B$.

    Thus Map $(A, B) = \emptyset$.

    (b) Suppose either $A = \emptyset$ or $B \neq \emptyset$.

    If $A = \emptyset$ then $A \times B = \emptyset$ and the only relation between $A$ and $B$ is the empty relation $\emptyset$ which has domain $A = \emptyset$ and is functional.

    Hence in this case $((A, B), \emptyset)$ is a mapping from $A$ to $B$ and so Map $(A, B) \neq \emptyset$.

    Finally, if $A \neq \emptyset$ and $B \neq \emptyset$ let $b$ be any element of $B$. Then $R = A \times \{b\}$ is a functional relation with domain $A$. So we see that $f = ((A, B), R)$ is a mapping from $A$ to $B$.

    Hence in this case also we have Map $(A, B) \neq \emptyset$.

28. Let $f = ((A, B), R)$, $g = ((B, C), S)$, $h = ((C, D), T)$. Then we have $g \circ f = ((A, C), S \circ R)$ and $h \circ g = ((B, D), T \circ S)$. Hence, using Exercise 17,

    $$h \circ (g \circ f) = ((A, D), T \circ (S \circ R)) = ((A, D), (T \circ S) \circ R) = (h \circ g) \circ f.$$

29. Let $a \in A$. Then we have $(f \circ I_A)(a) = f(I_A(a)) = f(a)$. Since Dom $(f \circ I_A) =$ Dom $f = A$ and the codomain of $f \circ I_A$ is the same as the codomain of $f$, it follows that $f \circ I_A = f$.

    Let Dom $g = B$; let $b$ be any element of $B$. Then $(I_A \circ g)(b) = I_A(g(b)) = g(b)$. Since Dom $(I_A \circ g) =$ Dom $g$ and the codomain of $I_A \circ g$ coincides with the codomain of $g$ $(= A)$, it follows that $I_A \circ g = g$.

30. (a) Suppose $f$ is injective. Let $a_0$ be any element of $A$.

    Let $S = \{z : ((\exists b)(\exists a)((b \in f^{\rightarrow}(A)) \wedge (z = (b, a)) \wedge (f(a) = b))) \vee ((\exists b)((b \notin f^{\rightarrow}(A)) \wedge (z = (b, a_0))))\}$.

    Then $S$ is a functional relation with domain $B$ and Range $S \subseteq A$.

    So $g = ((B, A), S)$ is a mapping from $B$ to $A$. Clearly $g \circ f = I_A$.

    (b) Suppose there is a mapping $g$ from $B$ to $A$ such that $g \circ f = I_A$.

    Let $X$ be any set and let $h_1$, $h_2$ be mappings from $X$ to $A$ such that $f \circ h_1 = f \circ h_2$. Then $g \circ (f \circ h_1) = g \circ (f \circ h_2)$; so $(g \circ f) \circ h_1 = (g \circ f) \circ h_2$, i.e. $I_A \circ h_1 = I_A \circ h_2$ and so $h_1 = h_2$.

(c) Suppose $f$ is not injective. Then there are elements $a_1$, $a_2$ of $A$ such that $a_1 \neq a_2$ but $f(a_1) = f(a_2)$.

Let $X = \{\emptyset\}$ and let $h_1$, $h_2$ be the mappings from $X$ to $A$ given by $h_1(\emptyset) = a_1$ and $h_2(\emptyset) = a_2$ respectively. Then $h_1 \neq h_2$ but $f \circ h_1 = f \circ h_2$. This is a contradiction to (3). Hence (3) $\Longrightarrow$ (1).

31. (1) Let $k_1$, $k_2$ be mappings from a set $X$ to $A$ such that $h \circ k_1 = h \circ k_2$, i.e. $(g \circ f) \circ k_1 = (g \circ f) \circ k_2$. Then $g \circ (f \circ k_1) = g \circ (f \circ k_2)$. Since $g$ is injective, $f \circ k_1 = f \circ k_2$. Since $f$ is injective, $k_1 = k_2$.

    So $h$ is injective.

    (2) Let $c$ be any element of $C$. Since $g$ is surjective there is an element $b$ of $B$ such that $g(b) = c$. Since $f$ is surjective there is an element $a$ of $A$ such that $f(a) = b$. Then $h(a) = g(f(a)) = g(b) = c$.

    So $h$ is surjective.

    (3) Let $k_1$, $k_2$ be mappings from a set $X$ to $A$ such that $f \circ k_1 = f \circ k_2$. Then $g \circ (f \circ k_1) = g \circ (f \circ k_2)$. Hence $(g \circ f) \circ k_1 = (g \circ f) \circ k_2$, i.e, $h \circ k_1 = h \circ k_2$. Since $h$ is injective it follows that $k_1 = k_2$.

    So $f$ is injective.

    (4) Let $c$ be any element of $C$.

    Since $h$ is surjective there is an element $a$ of $A$ such that $h(a) = c$. Then we have $c = h(a) = g(f(a))$.

    So $g$ is surjective.

    (5) Let $b_1$, $b_2$ be elements of $B$ such that $g(b_1) = g(b_2)$.

    Since $f$ is surjective there are elements $a_1$, $a_2$ of $A$ such that $b_1 = f(a_1)$ and $b_2 = f(a_2)$. Then $g(f(a_1)) = g(f(a_2))$, i.e. $h(a_1) = h(a_2)$.

    Since $h$ is injective, $a_1 = a_2$. Hence $b_1 = b_2$ and so $g$ is injective.

    (6) Let $b$ be an element of $B$.

    Then $g(b) \in C$ and since $h$ is surjective there is an element $a$ of $A$ such that $h(a) = g(b)$, i.e. $g(f(a)) = g(b)$.

    Since $g$ is injective it follows that $b = f(a)$. So $f$ is surjective.

32. Since $h \circ g \circ f$ is injective, $f$ is injective (by Exercise 31(3)).

    Since $g \circ f \circ h$ is injective, $h$ is injective (by Exercise 31(3)).

Since $f \circ h \circ g$ is surjective, $f$ is surjective (by Exercise 31(4)) and hence bijective.

Let $a$ be any element of $A$.

Then $f(a) \in B$ and since $f \circ h \circ g$ is surjective there is an element $b$ of $B$ such that $f(a) = (f \circ h \circ g)(b) = f(h(g(b)))$, Since $f$ is injective we have $a = h(g(b))$.

Thus $h$ is surjective and hence bijective.

Since $g \circ f \circ h$ is injective and $f \circ h$ is surjective (by Exercise 31(2)), $g$ is injective (by Exercise 31(5)).

Since $f \circ h \circ g$ is surjective and $f \circ h$ is injective (by Exercise 31(1)), $g$ is surjective (by Exercise 31(6)) and hence bijective.

33. (1) If $f^{\rightarrow}(\emptyset)$ were non-empty, containing an element $b$ say, then there would be an element $x$ of $\emptyset$ such that $f(x) = b$. But $\emptyset$ has no elements, so this is impossible.

Hence $f^{\rightarrow}(\emptyset) = \emptyset$.

(2) $y \in f^{\rightarrow}(\{x\}) \iff$ there is an element $t$ in $\{x\}$ such that $f(t) = y \iff y = f(x)$.

(3) Suppose $X_1 \subseteq X_2$.

Let $y$ be any element of $f^{\rightarrow}(X_1)$. Then there is an element $x_1$ of $X_1$ such that $f(x_1) = y$. Since $X_1 \subseteq X_2$ we have $x_1 \in X_2$. Thus $y = f(x_1) \in f^{\rightarrow}(X_2)$.

So $f^{\rightarrow}(X_1) \subseteq f^{\rightarrow}(X_2)$.

(4) Since $X_1 \subseteq X_1 \cup X_2$ and $X_2 \subseteq X_1 \cup X_2$ it follows from (3) above that $f^{\rightarrow}(X_1) \subseteq f^{\rightarrow}(X_1 \cup X_2)$ and $f^{\rightarrow}(X_2) \subseteq f^{\rightarrow}(X_1 \cup X_2)$. Hence $f^{\rightarrow}(X_1) \cup f^{\rightarrow}(X_2) \subseteq f^{\rightarrow}(X_1 \cup X_2)$.

If $y \in f^{\rightarrow}(X_1 \cup X_2)$ then there is an element $x$ of $X_1 \cup X_2$ such that $f(x) = y$. Now we have either $x \in X_1$ or $x \in X_2$ or both. Hence we have either $y = f(x) \in f^{\rightarrow}(X_1)$ or $y = f(x) \in f^{\rightarrow}(X_2)$ or both. Hence $y \in f^{\rightarrow}(X_1) \cup f^{\rightarrow}(X_2)$. So $f^{\rightarrow}(X_1 \cup X_2) \subseteq f^{\rightarrow}(X_1) \cup f^{\rightarrow}(X_2)$.

Hence $f^{\rightarrow}(X_1 \cup X_2) = f^{\rightarrow}(X_1) \cup f^{\rightarrow}(X_2)$.

(5) Since $X_1 \cap X_2 \subseteq X_1$ and $X_1 \cap X_2 \subseteq X_2$ it follows, using (3) above, that $f^{\rightarrow}(X_1 \cap X_2) \subseteq f^{\rightarrow}(X_1)$ and $f^{\rightarrow}(X_1 \cap X_2) \subseteq f^{\rightarrow}(X_2)$. Hence $f^{\rightarrow}(X_1 \cap X_2) \subseteq f^{\rightarrow}(X_1) \cap f^{\rightarrow}(X_2)$.

(6) Suppose $f$ is injective.

Let $y \in f^{\rightarrow}(X_1) \cap f^{\rightarrow}(X_2)$. Then there are elements $x_1$ of $X_1$ and $x_2$ of $X_2$ such that $f(x_1) = y$ and $f(x_2) = y$. Since $f$ is injective it follows that $x_1 = x_2$ and hence $x_1 = x_2 \in X_1 \cap X_2$. Thus $y \in f^{\rightarrow}(X_1 \cap X_2)$. Hence $f^{\rightarrow}(X_1) \cap f^{\rightarrow}(X_2) \subseteq f^{\rightarrow}(X_1 \cap X_2)$.

Combining this result with (5) above we have $f^{\rightarrow}(X_1) \cap f^{\rightarrow}(X_2) = f^{\rightarrow}(X_1 \cap X_2)$.

(7) We have $x \in f^{\leftarrow}(B \sim Y)$

$$\begin{aligned} &\iff f(x) \in B \sim Y \\ &\iff f(x) \notin Y \\ &\iff x \notin f^{\leftarrow}(Y) \iff x \in A \sim f^{\leftarrow}(Y). \end{aligned}$$

(8) Suppose $Y_1 \subseteq Y_2$.

If $x \in f^{\leftarrow}(Y_1)$ then we have $f(x) \in Y_1$ and so $f(x) \in Y_2$. Thus $x \in f^{\leftarrow}(Y_2)$.

So $f^{\leftarrow}(Y_1) \subseteq f^{\leftarrow}(Y_2)$.

(9) Since $Y_1 \subseteq Y_1 \cup Y_2$ and $Y_2 \subseteq Y_1 \cup Y_2$ it follows from (8) above that $f^{\leftarrow}(Y_1) \subseteq f^{\leftarrow}(Y_1 \cup Y_2)$ and $f^{\leftarrow}(Y_2) \subseteq f^{\leftarrow}(Y_1 \cup Y_2)$. So $f^{\leftarrow}(Y_1) \cup f^{\leftarrow}(Y_2) \subseteq f^{\leftarrow}(Y_1 \cup Y_2)$.

Let $x$ be any element of $f^{\leftarrow}(Y_1 \cup Y_2)$. Then $f(x) \in Y_1 \cup Y_2$. So we have $f(x) \in Y_1$ or $f(x) \in Y_2$ or both, i.e. $x \in f^{\leftarrow}(Y_1)$ or $x \in f^{\leftarrow}(Y_2)$ or both. Thus $x \in f^{\leftarrow}(Y_1) \cup f^{\leftarrow}(Y_2)$. Hence $f^{\leftarrow}(Y_1 \cup Y_2) \subseteq f^{\leftarrow}(Y_1) \cup f^{\leftarrow}(Y_2)$.

So $f^{\leftarrow}(Y_1 \cup Y_2) = f^{\leftarrow}(Y_1) \cup f^{\leftarrow}(Y_2)$.

(10) Since $Y_1 \cap Y_2 \subseteq Y_1$ and $Y_1 \cap Y_2 \subseteq Y_2$ it follows from (8) that $f^{\leftarrow}(Y_1 \cap Y_2) \subseteq f^{\leftarrow}(Y_1)$ and $f^{\leftarrow}(Y_1 \cap Y_2) \subseteq f^{\leftarrow}(Y_2)$; hence $f^{\leftarrow}(Y_1 \cap Y_2) \subseteq f^{\leftarrow}(Y_1) \cap f^{\leftarrow}(Y_2)$.

Conversely, suppose $x \in f^{\leftarrow}(Y_1) \cap f^{\leftarrow}(Y_2)$. Then $f(x) \in Y_1 \cap Y_2$ and so $x \in f^{\leftarrow}(Y_1 \cap Y_2)$. Hence $f^{\leftarrow}(Y_1) \cap f^{\leftarrow}(Y_2) \subseteq f^{\leftarrow}(Y_1 \cap Y_2)$.

So $f^{\leftarrow}(Y_1) \cap f^{\leftarrow}(Y_2) = f^{\leftarrow}(Y_1 \cap Y_2)$.

(11) Suppose $f$ is bijective.

Let $y \in f^{\rightarrow}(A \sim X)$. Then there is an element $x'$ of $A$ not in $X$ such that $y = f(x')$. If $y$ were in $f^{\rightarrow}(X)$ there would be an element $x$ of $X$ such that $y = f(x)$ and hence, since $f$ is injective, we would have $x\ (\in X) = x'\ (\notin X)$, which is a contradiction. Hence we have $f^{\rightarrow}(A \sim X) \subseteq B \sim f^{\rightarrow}(X)$.

Conversely, suppose $y \in B \sim f^{\rightarrow}(X)$. Since $f$ is surjective there is an element $a$ of $A$ such that $f(a) = y$. Since $y \notin f^{\rightarrow}(X)$ we must have $a \notin X$, i.e. $a \in A \sim X$ and so $y \in f^{\rightarrow}(A \sim X)$. Thus $B \sim f^{\rightarrow}(X) \subseteq f^{\rightarrow}(A \sim X)$.

(12) Let $x$ be any element of $X$.

Then $f(x) \in f^{\rightarrow}(X)$ and hence $x \in f^{\leftarrow}(f^{\rightarrow}(X))$. So we have $X \subseteq f^{\leftarrow}(f^{\rightarrow}(X))$.

(13)(a) Suppose $f$ is injective.

Let $X$ be any subset of $A$ and let $t$ be any element of $f^{\leftarrow}(f^{\rightarrow}(X))$. Then $f(t) \in f^{\rightarrow}(X)$. Hence there is an element $x$ of $X$ such that $f(t) = f(x)$. Since $f$ is injective it follows that $t = x$, i.e. $t \in X$. So $f^{\leftarrow}(f^{\rightarrow}(X)) \subseteq X$, and hence, using (12), we have $X = f^{\leftarrow}(f^{\rightarrow}(X))$.

(b) Conversely, suppose that for every subset $X$ of $A$ we have $X = f^{\leftarrow}(f^{\rightarrow}(X))$. Let $a_1$, $a_2$ be elements of $A$ such that $f(a_1) = f(a_2)$. Let $X = \{a_1\}$. Then $f^{\rightarrow}(X) = \{f(a_1)\}$ and so we have $a_2 \in f^{\leftarrow}(f^{\rightarrow}(X)) = X = \{a_1\}$. Thus $a_2 = a_1$.

Thus $f$ is injective.

(14) Let $b$ be any element of $f^{\rightarrow}(f^{\leftarrow}(Y))$.

Then there is an element $a$ of $f^{\leftarrow}(Y)$ such that $b = f(a)$. Since $a \in f^{\leftarrow}(Y)$ it follows that $f(a) \in Y$, i.e. $b \in Y$.

Hence $f^{\rightarrow}(f^{\leftarrow}(Y)) \subseteq Y$.

(15)(a) Suppose $f$ is surjective.

Let $y$ be any element of $Y$. There is an element $a$ of $A$ such that $y = f(a)$. Since $f(a) \in Y$ we have $a \in f^{\leftarrow}(Y)$ and hence $y = f(a) \in f^{\rightarrow}(f^{\leftarrow}(Y))$. Thus $Y \subseteq f^{\rightarrow}(f^{\leftarrow}(Y))$ and hence, using (14), we have $Y = f^{\rightarrow}(f^{\leftarrow}(Y))$.

(b) Conversely, suppose that for every subset $Y$ of $B$ we have $f^{\rightarrow}(f^{\leftarrow}(Y)) = Y$.

Let $b$ be any element of $B$. Take $Y = \{b\}$. Then we have $\{b\} = f^{\rightarrow}(f^{\leftarrow}(\{b\})) \subseteq f^{\rightarrow}(A)$. Hence $b \in f^{\rightarrow}(A)$.

So $f$ is surjective.

34. (1) Suppose there exists a mapping $h$ from $A$ to $B$ such that $f = g \circ h$.

    Let $c$ be any element of $f^{\rightarrow}(A)$. Then there is an element $a$ of $A$ such that $c = f(a)$. Then $c = f(a) = g(h(a)) \in g^{\rightarrow}(B)$. So $f^{\rightarrow}(A) \subseteq g^{\rightarrow}(B)$.

    (2) Conversely, suppose that$f^{\rightarrow}(A) \subseteq g^{\rightarrow}(B)$.

    Since $g$ is injective there is a mapping $k$ from $C$ to $B$ such that $k \circ g = I_B$. Let $h = k \circ f$. Let $a$ be any element of $A$. Then there is an element $b$ of $B$ such that $f(a) = g(b)$. It follows that $(g \circ h)(a) = g(k(f(a))) = g(k(g(b))) = g(b) = f(a)$.

    So $g \circ h = f$.

    If $g \circ h_1 = f = g \circ h$ we have $h_1 = h$ since $g$ is injective.

35. Let $f$ be defined by setting $f(x) = (f_1(x), f_2(x))$ for all $x$ in $X$.

    Then $(\pi_1 \circ f)(x) = f_1(x)$ and $(\pi_2 \circ f)(x) = f_2(x)$ for all $x$ in $X$. So $\pi_1 \circ f = f_1$ and $\pi_2 \circ f = f_2$.

    Let $g$ be another mapping from $X$ to $A_1 \times A_2$ such that $\pi_1 \circ g = f_1$ and $\pi_2 \circ g = f_2$. Then for every element $x$ of $X$ we have

    first coordinate of $g(x) = \pi_1(g(x)) = f_1(x)$ and
    second coordinate of $g(x) = \pi_2(g(x)) = f_2(x)$.

    So $g(x) = (f_1(x), f_2(x)) = f(x)$ for all elements $x$ of $X$. Thus $g = f$, i.e. $f$ is unique.

36. The natural way to define a mapping $f$ $(= f_1 \times f_2)$ from $A_1 \times A_2$ to $B_1 \times B_2$ is to set $f((a_1, a_2)) = (f(a_1), f_2(a_2))$ for all ordered pairs $(a_1, a_2)$ in $A_1 \times A_2$.

    Suppose $f_1$ and $f_2$ are bijections.

    If $(f_1 \times f_2)((a_1, a_2)) = (f_1 \times f_2)((a_1', a_2'))$ then $(f_1(a_1), f_2(a_2)) = (f(a_1'), f_2(a_2'))$, whence $f_1(a_1) = f_1(a_1')$ and $f_2(a_2) = f_2(a_2')$. Since $f_1$ and $f_2$ are injective it follows that $a_1 = a_1'$ and $a_2 = a_2'$, i.e. $(a_1, a_2) = (a_1', a_2')$. So $f_1 \times f_2$ is injective.

    Let $(b_1, b_2)$ be any element of $B_1 \times B_2$. Since $f_1$ and $f_2$ are surjective there are elements $a_1$ of $A_1$ and $a_2$ of $A_2$ such that $f_1(a_1) = b_1$ and $f_2(a_2) = b_2$. Then $(b_1, b_2) = (f_1(a_1), f_2(a_2)) = (f_1 \times f_2)((a_1, a_2))$. So $f_1 \times f_2$ is surjective.

37. The natural way to define a mapping $h = \text{Map}(f,g)$ from Map $(A_1, B_1)$ to Map $(A_2, B_2)$ is to set $h(k) = g \circ k \circ f$ for all mappings $k$ from $A_1$ to $B_1$.

38. In order to define a mapping $F$ from Map $(A,$ Map $(B,C))$ to Map $(B \times A, C)$ we set for each mapping $f$ in Map$(A,$ Map $(B,C))$

$$(F(f))((b,a)) = (f(a))(b) \text{ for all } (b,a) \text{ in } B \times A.$$

$F$ is injective. For suppose $F(f_1) = F(f_2)$ for elements $f_1$, $f_2$ of Map $(A,$ Map $(B,C))$. Then for all $(b,a)$ in $B \times A$ we have $(f_1(a))(b) = (f_2(a))(b)$. So $f_1(a) = f_2(a)$ for all $a$ in $A$. Thus $f_1 = f_2$.

$F$ is surjective. For let $g$ be any element of Map $(B \times A, C)$. Define a mapping $f$ from $A$ to Map $(B,C)$ by setting for each $a$ in $A$

$$(f(a))(b) = g(b,a) \text{ for all } b \text{ in } B.$$

Then $F(f) = g$.

39. The natural way to define a mapping $F$ from Map $(C, A \times B)$ to Map $(C,A)\times$ Map $(C,B)$ is to define for each mapping $f$ in Map $(C, A \times B)$

$$F(f) = (\pi_1 \circ f, \pi_2 \circ f)$$

where $\pi_1$ and $\pi_2$ are the first and second projections from $A \times B$ to $A$ and $B$ respectively.

To show that $F$ is injective suppose $F(f_1) = F(f_2)$ for $f_1$, $f_2$ in Map $(C, A \times B)$. Then $(\pi_1 \circ f_1, \pi_2 \circ f_1) = (\pi_1 \circ f_2, \pi_2 \circ f_2)$; so $\pi_1 \circ f_1 = \pi_1 \circ f_2$ and $\pi_2 \circ f_1 = \pi_2 \circ f_2$. Thus for every element $c$ of $C$ we have $\pi_1(f_1(c)) = \pi_1(f_2(c))$ and $\pi_2(f_1(c)) = \pi_2(f_2(c))$. So for every element $c$ of $C$ we have $f_1(c) = f_2(c)$. Thus $f_1 = f_2$.

$F$ is surjective. For suppose $(g_1, g_2) \in$ Map $(C,A)\times$ Map $(C,B)$. Let $f$ be the mapping from $C$ to $A \times B$ defined by setting

$$f(c) = (g_1(c), g_2(c)) \text{ for all } c \text{ in } C.$$

Then $F(f) = (g_1, g_2)$.

# Chapter 17

# ANSWERS TO CHAPTER 4

40. (1) Let $x \in \bigcup_{k \in K} X_{f(k)}$. Then there is an index $k_1$ in $K$ such that $x \in X_{f(k_1)}$. Since $f(k_1) \in I$ it follows that $x \in \bigcup_{i \in I} X_i$. So $\bigcup_{k \in K} X_{f(k)} \subseteq \bigcup_{i \in I} X_i$.

    Conversely, let $x \in \bigcup_{i \in I} X_i$. Then there is an index $i_1$ in $I$ such that $x \in X_{i_1}$. Since $f$ is a surjection there is an index $k_1$ in $K$ such that $f(k_1) = i_1$. Thus $x \in X_{f(k_1)}$ and so we have $x \in \bigcup_{k \in K} X_{f(k)}$. Hence $\bigcup_{i \in I} X_i \subseteq \bigcup_{k \in K} X_{f(k)}$.

    Thus $\bigcup_{i \in I} X_i = \bigcup_{k \in K} X_{f(k)}$.

    (2) Let $x \in \bigcap_{k \in K} X_{f(k)}$. Let $i_1$ be any index in $I$. Since $f$ is surjective there is an index $k_1$ in $K$ such that $f(k_1) = i_1$. Thus $x \in X_{f(k_1)} = X_{i_1}$. So $x \in \bigcap_{i \in I} X_i$. Hence $\bigcap_{k \in K} X_{f(k)} \subseteq \bigcap_{i \in I} X_i$.

    Conversely, let $x \in \bigcap_{i \in I} X_i$. Let $k_1$ be any index in $K$. Then $f(k_1) \in I$ and so $x \in X_{f(k_1)}$. Hence $x \in \bigcap_{k \in K} X_{f(k)}$. So we have $\bigcap_{i \in I} X_i \subseteq \bigcap_{k \in K} X_{f(k)}$.

    Hence $\bigcap_{i \in I} X_i = \bigcap_{k \in K} X_{f(k)}$.

41. (1) Let $x \in \bigcup_{i \in I} X_i$. Then there is an index $i_1$ in $I$ such that $x \in X_{i_1}$. Since $I = \bigcup_{k \in K} J_k$ there is an index $k_1$ in $K$ such that $i_1 \in J_{k_1}$. So $x \in \bigcup_{i \in J_{k_1}} X_i$ and hence $x \in \bigcup_{k \in K} \left( \bigcup_{i \in J_k} X_i \right)$. So $\bigcup_{i \in I} X_i \subseteq \bigcup_{k \in K} \left( \bigcup_{i \in J_k} X_i \right)$.

Conversely, suppose $x \in \bigcup_{k \in K}\left(\bigcup_{i \in J_k} X_i\right)$. Then there is an index $k_1$ in $K$ such that $x \in \bigcup_{i \in J_{k_1}} X_i$. Hence there is an index $i_1$ in $J_{k_1}$ such that $x \in X_{i_1}$. Since $J_{k_1} \subseteq I$ it follows that we have $x \in \bigcup_{i \in I} X_i$. So $\bigcup_{k \in K}\left(\bigcup_{i \in J_k} X_i\right) \subseteq \bigcup_{i \in I} X_i$.

Thus $\bigcup_{i \in I} X_i = \bigcup_{k \in K}\left(\bigcup_{i \in J_k} X_i\right)$.

(2) Let $x \in \bigcap_{i \in I} X_i$. Let $k_1$ be any index in $K$. Then for every index $i$ in $I$, and hence for every index $i$ in $J_{k_1}$ we have that $x \in X_i$. So $x \in \bigcap_{I \in J_{k_1}} X_i$ and hence $x \in \bigcap_{k \in K}\left(\bigcap_{i \in J_k} X_i\right)$. It follows that $\bigcap_{i \in I} X_i \subseteq \bigcap_{k \in K}\left(\bigcap_{i \in J_k} X_i\right)$.

Conversely, suppose $x \in \bigcap_{k \in K}\left(\bigcap_{i \in J_k} X_i\right)$. Let $i_1$ be any index in $I$. Then, since $I = \bigcup_{k \in K} J_k$, there is an index $k_1$ in $K$ such that $i_1 \in J_{k_1}$. Since $x \in \bigcap_{i \in J_{k_1}} X_i$ it follows that $x \in X_{i_1}$. So $x \in \bigcap_{i \in I} X_i$. Hence $\bigcap_{k \in K}\left(\bigcap_{i \in J_k} X_i\right) \subseteq \bigcap_{i \in I} X_i$.

Thus $\bigcap_{i \in I} X_i = \bigcap_{k \in K}\left(\bigcap_{i \in J_k} X_i\right)$.

42. Suppose $(a, c) \in S \circ \left(\bigcup_{i \in I} R_i\right)$. Then there is an element $b$ of $B$ such that $(a, b) \in \bigcup_{i \in I} R_i$ and $(b, c) \in S$. There is then an index $i_1$ in $I$ such that $(a, b) \in R_{i_1}$. Since we have $(a, b) \in R_{i_1}$ and $(b, c) \in S$ we have $(a, c) \in S \circ R_{i_1}$ and so $(a, c) \in \bigcup_{i \in I} S \circ R_i$. Thus $S \circ \left(\bigcup_{i \in I} R_i\right) \subseteq \bigcup_{i \in I} S \circ R_i$.

    Conversely, suppose $(a, c) \in \bigcup_{i \in I} S \circ R_i$. Then there is an index $i_1$ in $I$ such that $(a, c) \in S \circ R_{i_1}$. There is then an element $b$ of $B$ such that $(a, b) \in R_{i_1}$ and $(b, c) \in S$. Since $(a, b) \in R_{i_1}$ we have $(a, b) \in \bigcup_{i \in I} R_i$. Hence $(a, c) \in S \circ \left(\bigcup_{i \in I} R_i\right)$. So $\bigcup_{i \in I} S \circ R_i \subseteq S \circ \left(\bigcup_{i \in I} R_i\right)$.

    Hence $S \circ \left(\bigcup_{i \in I} R_i\right) = \bigcup_{i \in I} S \circ R_i$.

43. Consider the relation $F =$

$$\{(x, y) : (x \in D) \wedge (\exists i)((i \in I) \wedge (x \in \operatorname{Dom} F_i) \wedge (((x, y) \in F_i)))\}.$$

    Then $F$ has domain $D$. We claim that $F$ is functional. So suppose that for some element $x$ of $D$ we have $(x, y) \in F$ and $(x, y') \in F$. Then there are indices $i$, $j$ in $I$ such that

$$x \in \operatorname{Dom} F_i \text{ and } (x, y) \in F_i$$

and

$$x \in \operatorname{Dom} F_j \text{ and } (x, y) \in F_j.$$

Then $x \in \operatorname{Dom} F_i \cap \operatorname{Dom} F_j$ and so $y = F_i(x) = F_j(x) = y'$. So $F$ is functional.

For every element $x$ of $\operatorname{Dom} F_i$ we have $(x, F_i(x)) \in F_i$ and so $(x, F_i(x)) \in F$; hence $F(x) = F_i(x)$.

44. For each index $i$ in $I$ let $f_i = ((A_i, B), F_i)$. Then the family $(F_i)_{i \in I}$ of functional relations satisfies the condition of Exercise 43. Hence there is a functional relation $F$ with domain $A$ such that for every index $i$ in $I$ and every element $x$ of $A_i$ we have $F(x) = F_i(x)$. If we set $f = ((A, B), F)$ it follows that for every index $i$ in $I$ and every element $x$ of $A_i$ we have $f(x) = f_i(x)$.

45. For each element $a$ of $A$ let $f_a$ be the mapping from $I$ to $\bigcup_{i \in I} X_i$ given by $f_a(i) = f_i(a)$ for all indices $i$ in $I$. The mapping $f$ defined by setting $f(a) = f_a$ for all elements $a$ of $A$ maps $A$ to $P$. Then for each index $i$ in $I$ and each element $a$ of $A$ we have $(\pi_i \circ f)(a) = \pi_i(f_a) = f_i(a)$. So $\pi_i \circ f = f_i$ for each $i$ in $I$.

    To show that $f$ is the only mapping with this property suppose we have another mapping $f'$ from $A$ to $P$ such that $\pi_i \circ f' = f_i$ for all indices $i$ in $I$. Then for all elements $a$ of $A$ and all indices $i$ in $I$ we have $(f(a))(i) = f_i(a) = (\pi_i \circ f')(a) = (f'(a))(i)$. So for all elements $a$ of $A$ we have $f(a) = f'(a)$. Hence $f = f'$.

46. Let $x \in \bigcup_{k \in K} \left(\bigcap_{i \in J_k} X_{i,k}\right)$. Let $f_1$ be any element of $I$. There is an index $k_0$ in $K$ such that $x \in \bigcap_{i \in J_{k_0}} X_{i,k_0}$. Thus $x$ is in every one of the sets $X_{i,k_0}$ with $i$ in $J_{k_0}$; hence, in particular, we have $x \in X_{f_1(k_0),k_0}$ and so $x \in \bigcup_{k \in K} X_{f_1(k),k}$ for every $f_1$ in $I$. Thus it follows that $x \in \bigcap_{f \in I} \left(\bigcup_{k \in K} X_{f(k),k}\right)$. So $\bigcup_{k \in K} \left(\bigcap_{i \in J_k} X_{i,k}\right) \subseteq \bigcap_{f \in I} \left(\bigcup_{k \in K} X_{f(k),k}\right)$.

    Conversely, suppose $x \notin \bigcup_{k \in K} \left(\bigcap_{i \in J_k} X_{i,k}\right)$.

    Then for every index $k$ in $K$ we have $x \notin \bigcap_{i \in J_k} X_{i,k}$. Hence for every index $k$ in $K$ there is an index $j_k$ in $J_k$ such that $x \notin X_{j_k,k}$. Let $f_0$ be the element of $\prod_{k \in K} J_k$ given by $f_0(k) = j_k$ for each index $k$ in $K$. Then for each $k$ in $K$ we have $x \notin X_{f_0(k),k}$; hence

$x \notin \bigcup_{k\in K} X_{f_0(k),k}$. So finally $x \notin \bigcap_{f\in I}\left(\bigcup_{k\in K} X_{f(k),k}\right)$. Consequently $\bigcap_{f\in I}\left(\bigcup_{k\in K} X_{f(k),k}\right) \subseteq \bigcup_{k\in K}\left(\bigcap_{i\in J_k} X_{i,k}\right)$.

Thus $\bigcup_{k\in K}\left(\bigcap_{i\in J_k} X_{i,k}\right) = \bigcap_{f\in I}\left(\bigcup_{k\in K} X_{f(k),k}\right)$.

Suppose now $K = \{1,2\}$, $J_1 = \{1\}$, $J_2 = \{1,2\}$.

Then $I = \prod_{i\in K} J_k = \{f_1, f_2\}$ where $f_1(1) = 1$, $f_1(2) = 1$ and $f_2(1) = 1$, $f_2(2) = 2$. Hence we have $\bigcap_{i\in J_1} X_{i,1} = X_{1,1}$ and $\bigcap_{i\in J_2} X_{i,2} = X_{1,2} \cap X_{2,2}$. So

$$\bigcup_{k\in K}\left(\bigcap_{i\in J_k} X_{i,k}\right) = X_{1,1} \cup (X_{1,2} \cap X_{2,2}),$$

$$\bigcup_{k\in K} X_{f_1(k),k} = X_{1,1} \cup X_{1,2},$$

$$\bigcup_{k\in K} X_{f_2(k),k} = X_{1,1} \cup X_{2,2}.$$

Hence

$$\bigcap_{f\in I}\left(\bigcup_{k\in K} X_{f(k),k}\right) = (X_{1,1} \cup X_{1,2}) \cap (X_{1,1} \cup X_{2,2}).$$

Thus we have

$$X_{1,1} \cup (X_{1,2} \cap X_{2,2}) = (X_{1,1} \cup X_{1,2}) \cap (X_{1,1} \cup X_{2,2})$$

and so in this case the result of Exercise 46 specialises to the distributive property.

47. (a) Suppose $f$ is surjective.

Then for every element $b$ of $B$ the set $f^{\leftarrow}(\{b\})$ is non-empty. By the Axiom of Choice there exists a choice function for the family $(f^{\leftarrow}(\{b\}))_{b\in B}$, i.e. a mapping $g$ from $B$ to $\bigcup_{b\in B} f^{\leftarrow}(\{b\}) = A$ such that $g(b) \in f^{\leftarrow}(\{b\})$ for all elements $b$ of $B$. So for each element $b$ of $B$ we have $f(g(b)) = b$, and hence $f \circ g = I_B$.

(b) Suppose there is a mapping $g$ from $B$ to $A$ such that $f \circ g = I_B$.

Let $Y$ be any non-empty set and let $h_1$, $h_2$ be mappings from $B$ to $Y$ such that $h_1 \circ f = h_2 \circ f$. Then we have $(h_1 \circ f) \circ g = (h_2 \circ f) \circ g$, whence $h_1 \circ (f \circ g) = h_2 \circ (f \circ g)$, i.e. $h_1 \circ I_B = h_2 \circ I_B$ and so $h_1 = h_2$.

(c) Suppose $f$ were not surjective; then there would be an element $b_0$ of $B$ not in the range of $f$. Let $Y = \{1, 2\}$. Let $h_1$ be the mapping from $B$ to $Y$ given by

$$h_1(b) = 1 \text{ for all elements } b \text{ of } B;$$

let $h_2$ be the mapping from $B$ to $Y$ given by

$$h_2(b) = \begin{cases} 1 & \text{for all elements } b \text{ in the range of } f, \\ 2 & \text{for all elements } b \text{ not in the range of } f. \end{cases}$$

Then $h_1 \circ f = h_2 \circ f$ but $h_1 \neq h_2$ (since $h(b_0) = 1$ but $h_2(b_0) = 2$), and this contradicts (3).

48. Consider the family $(x)_{x \in X}$. The sets of this family are all non-empty. So there exists a choice function $s$, i.e. a mapping $s$ from $X$ to $\bigcup_{x \in X} x = \bigcup X$ such that for each set $x$ in $X$ we have $s(x) \in x$.

    If $s(x_1) = s(x_2)$ we have $s(x_1) \in x_1 \cap x_2$, whence $x_1 = x_2$. Thus $s$ is an injection.

49. Consider the family $(R^{\rightarrow}(\{a\}))_{a \in A}$; this is a family of non-empty sets and so there is a choice function $s$, i.e. a mapping $s$ from $A$ to $\bigcup_{a \in A} R^{\rightarrow}(\{a\}) \subseteq B$ such that $s(a) \in R^{\rightarrow}(\{a\})$ for each $a$ in $A$. Let $F$ be the set of all pairs of the form $(a, s(a))$ with $a \in A$.

    Then $((A, B), F)$ is a mapping from $A$ to $B$ such that $F \subseteq R$.

50. $(f_i \circ \pi_i)_{i \in I}$ is a family of mappings from $P$ to $(A'_i)_{i \in I}$. Hence there is a mapping $f$ from $P$ to $P'$ such that $\pi'_i \circ f = f_i \circ \pi_i$ for all indices $i$ in $I$.

51. Suppose all the mappings $f_i$ are injective.

    Then for each index $i$ in $I$ there is a mapping $g_i$ from $A'_i$ to $A_i$ such that $g_i \circ f_i = I_{A_i}$. Since $(g_i \circ \pi'_i)_{i \in I}$ is a family of mappings from $P'$ to $(A_i)_{i \in I}$ there is a unique mapping $g$ from $P'$ to $P$ such that $\pi_i \circ g = g_i \circ \pi'_i$ for all indices $i$ in $I$. Then we have

$$\pi_i \circ g \circ f = g_i \circ \pi'_i \circ f = g_i \circ f_i \circ \pi_i = \pi_i = \pi_i \circ I_P$$

    for all indices $i$ in $I$.

    So $g \circ f = I_P$ and hence $f$ is injective.

52. Define a mapping $F$ from $\prod_{i \in I} X_i$ to $X_1 \times X_2$ by setting

$$F(f) = (f(1), f(2)) \text{ for each } f \text{ in } \prod_{i \in I} X_i$$

and a mapping $G$ from $X_1 \times X_2$ to $\prod_{i \in I} X_i$ by setting

$$\begin{aligned} (G(x))(1) &= x_1 \\ (G(x))(2) &= x_2 \end{aligned}$$

for each element $x = (x_1, x_2)$ in $X_1 \times X_2$.

Then it is easily verified that $F \circ G$ and $G \circ F$ are the identity mappings of $X_1 \times X_2$ and $\prod_{i \in I} X_i$ respectively. Namely, for each element $x = (x_1, x_2)$ of $X_1 \times X_2$ we have

$$(F \circ G)(x) = F(G(x)) = ((G(x))(1), (G(x))(2)) = (x_1, x_2) = x$$

while for each mapping $f$ in $\prod_{i \in I} X_i$ we have

$$((G \circ F)(f))(1) = (G(F(f)))(1) = (F(f))_1 = f(1)$$

and similarly

$$((G \circ F)(f))(2) = f(2).$$

Thus $(G \circ F)(f) = f$.

So $F$ is a bijection from $\prod_{i \in I} X_i$ to $X_1 \times X_2$.

# Chapter 18

# ANSWERS TO CHAPTER 5

53. (1) Let $(a, b) \in R \cup R^{-1}$. If $(a, b) \in R$ then $(b, a) \in R^{-1}$ and so $(b, a) \in R \cup R^{-1}$. Similarly if $(a, b) \in R^{-1}$ then $(b, a) \in R \cup R^{-1}$. So $R \cup R^{-1}$ is symmetric. Of course $R \cup R^{-1} \supseteq R$.

    So $R \cup R^{-1}$ is a symmetric relation including $R$.

    Let $S$ be any symmetric relation which includes $R$. Then if $(a, b) \in R^{-1}$ we have $(b, a) \in R$; so $(b, a) \in S$ and hence $(a, b) \in S$. So $R^{-1} \subseteq S$. Thus $S \supseteq R \cup R^{-1}$.

    So $R \cup R^{-1}$ is the smallest symmetric relation including $R$.

    (2) Let $(a, b) \in R \cap R^{-1}$. Then $(a, b) \in R$ and $(a, b) \in R^{-1}$. So $(b, a) \in R^{-1}$ and $(b, a) \in R$. Thus $(b, a) \in R \cap R^{-1}$. So $R \cap R^{-1}$ is symmetric and of course $R \cap R^{-1} \subseteq R$.

    Let $T$ be any symmetric relation included in $R$. Let $(a, b) \in T$. Then $(b, a) \in T$ also. So $(a, b) \in R$ and $(b, a) \in R$, i.e. $(a, b) \in R$ and $(a, b) \in R^{-1}$. Thus $(a, b) \in R \cap R^{-1}$. So $T \subseteq R \cap R^{-1}$.

    So $R \cap R^{-1}$ is the largest symmetric relation included in $R$.

54. Let $(a, c) \in R \circ R$. Then there is an element $b$ of $X$ such that $(a, b) \in R$ and $(b, c) \in R$. Since $R$ is transitive it follows that $(a, c) \in R$. So $R \circ R \subseteq R$.

    Conversely, suppose $(a, b) \in R$. Since $R$ is reflexive we have $(a, a) \in R$. Hence $(a, b) \in R \circ R$. Thus $R \subseteq R \circ R$.

    Hence $R \circ R = R$.

55. (1) Suppose $R$ is an equivalence relation.

    Since $R$ is reflexive we have $D_X \subseteq R$.

    Let $(a,d) \in R \circ R^{-1} \circ R$. Then there are elements $b$, $c$ of $X$ such that $(a,b) \in R$, $(b,c) \in R^{-1}$, $(c,d) \in R$. Since $R$ is symmetric, $R^{-1} = R$ and so $(b,c) \in R$. Then, since $R$ is transitive, we have $(a,d) \in R$. So $R \circ R^{-1} \circ R \subseteq R$. Conversely, suppose $(a,d) \in R$. Since $(d,d) \in R^{-1}$ and $(d,d) \in R$ we have $(a,d) \in R \circ R^{-1} \circ R$. Thus $R \subseteq R \circ R^{-1} \circ R$.

    So $R \circ R^{-1} \circ R = R$.

    (2) Suppose $R \supseteq D_X$ and $R \circ R^{-1} \circ R = R$.

    Since $R \supseteq D_X$, $R$ is reflexive.

    Suppose $(a,b) \in R$. Then $(b,a) \in R^{-1}$. Since $(a,a) \in R$ and $(b,b) \in R$ we have $(b,a) \in R \circ R^{-1} \circ R = R$. So $R$ is symmetric.

    Suppose $(a,b) \in R$ and $(b,c) \in R$. Since $(b,b) \in R$ we have $(b,b) \in R^{-1}$. So $(a,c) \in R \circ R^{-1} \circ R = R$. Thus $R$ is transitive.

    Hence $R$ is an equivalence relation.

56. Since $R$ and $S$ are equivalence relations we have $D_X \subseteq R$, $D_X \subseteq S$, $R = R^{-1}$, $S = S^{-1}$, $R \circ R = R$, $S \circ S = S$.

    (1) Suppose $R \circ S$ is an equivalence relation. Then we have $S \circ R = S^{-1} \circ R^{-1} = (R \circ S)^{-1} = R \circ S$.

    (2) Conversely, suppose $R \circ S = S \circ R$.

    Since $D_X \subseteq D_X \circ D_X \subseteq R \circ S$, we see that $R \circ S$ is reflexive.

    Since $(R \circ S)^{-1} = S^{-1} \circ R^{-1} = S \circ R = R \circ S$ we have that $R \circ S$ is symmetric.

    Finally, since $(R \circ S) \circ (R \circ S) = R \circ (S \circ R) \circ S = R \circ (R \circ S) \circ S = (R \circ R) \circ (S \circ S) \subseteq R \circ S$, it follows that $R \circ S$ is transitive.

    It is easy to verify that $R \circ S \supseteq R \circ D_X = R$ and similarly $R \circ S \supseteq S$. So if $S \circ R = R \circ S$ then $R \circ S$ is an equivalence relation on $X$ which includes both $R$ and $S$.

    Let $T$ be any such relation. Then, using Exercise 54 we have that $T = T \circ T \supseteq R \circ S$. Thus $R \circ S$ is indeed the intersection of all equivalence relations on $X$ which include both $R$ and $S$.

57. (1) Let $(x, y) \in R$.

Since $\pi_1^{\rightarrow}(A) = E$, we have $x \in \pi_1^{\rightarrow}(A)$. So there is an element $t$ of $E$ such that $(x, t) \in A$. Since $R$ is symmetric we have $(t, x) \in R$. Since $R$ is transitive we deduce that $(t, y) \in R$. Since $(x, t) \in A$ and $(t, y) \in R$ it follows that $(x, y) \in R \circ A$.

So $R \subseteq R \circ A$.

Conversely, let $(x, y) \in R \circ A$.

Then there is an element $t$ of $E$ such that $(x, t) \in A$ and $(t, y) \in R$. Since $A \subseteq R$ we have $(x, t) \in R$ and $(t, y) \in R$; hence, since $R$ is transitive, we have $(x, y) \in R$.

So $R \circ A \subseteq R$ and hence $R \circ A = R$.

(2) Let $(x, y) \in (R \cap S) \circ A$.

Then there is an element $t$ of $E$ such that $(x, t) \in A$ and $(t, y) \in R \cap S$. Since $A \subseteq R$ and $R \cap S \subseteq R$ we have $(x, t) \in R$ and $(t, y) \in R$; hence $(x, y) \in R$. Since $(x, t) \in A$ and $(t, y) \in S$ we have $(x, y) \in S \circ A$.

Thus $(R \cap S) \circ A \subseteq R \cap (S \circ A)$.

Conversely, let $(x, y) \in R \cap (S \circ A)$. Then $(x, y) \in R$ and there is an element $t$ of $E$ such that $(x, t) \in A$ and $(t, y) \in S$. Since $(x, t) \in A$ we have $(x, t) \in R$ and so $(t, x) \in R$. Hence $(t, y) \in R$. Hence $(t, y) \in R \cap S$ and it follows that $(x, y) \in (R \cap S) \circ A$.

Thus $R \cap (S \circ A) \subseteq (R \cap S) \circ A$ and so $R \cap (S \circ A) = (R \cap S) \circ A$.

58. (1) Suppose $(a, b) \in R$.

Let $x$ be any element of $R(a)$. By definition of $R(a)$ we have $(a, x) \in R$. Since $(a, b) \in R$ and $R$ is symmetric we see that $(b, a) \in R$. Then, since $R$ is transitive, we deduce that $(b, x) \in R$, i.e. $x \in R(b)$. Thus $R(a) \subseteq R(b)$. Next, let $x$ be any element of $R(b)$, so that $(b, x) \in R$. Since $(a, b) \in R$ and $(b, x) \in R$ we have $(a, x) \in R$ since $R$ is transitive. So $x \in R(a)$ and hence $R(b) \subseteq R(a)$.

Thus $R(a) = R(b)$.

Conversely, suppose $R(a) = R(b)$.

Since $R$ is reflexive we have $(b, b) \in R$ and so $b \in R(b)$. Hence $b \in R(a)$ and so, by definition of $R(a)$, we have $(a, b) \in R$.

(2) Suppose $R(a) \neq R(b)$.

If $R(a)$ and $R(b)$ are not disjoint there is an element $t$ of $E$ such that $t \in R(a) \cap R(b)$. Then $(a,t) \in R$ and $(b,t) \in R$. Since $R$ is symmetric we have $(t,b) \in R$ and so, since $R$ is transitive, $(a,b) \in R$; from this it follows (using (1) above), that $R(a) = R(b)$, which is a contradiction.

59. (1) Let $f \in P$. Then $(\pi_i(f), \pi_i(f)) \in R_i$ for each index $i$; so $(f,f) \in R$.

Suppose $(f,g) \in R$. Then $(\pi_i(f), \pi_i(g)) \in R_i$ for each index $i$; hence $(\pi_i(g), \pi_i(f)) \in R_i$ for each index $i$. So $(g,f) \in R$.

Suppose $(f,g) \in R$ and $(g,h) \in R$. Then $(\pi_i(f), \pi_i(g)) \in R_i$ and $(\pi_i(g), \pi_i(h)) \in R_i$ for each index $i$; hence $(\pi_i(f), \pi_i(h)) \in R_i$ for each index $i$. So $(f,h) \in R$.

Thus $R$ is an equivalence relation.

(2) Let $\eta$ be the canonical surjection from $P$ onto $P/R$; for each index $i$ in $I$ let $\eta_i$ be the canonical surjection from $A_i$ onto $A_i/R_i$. Since $\eta$ and each $\eta_i$ are surjections there are mappings $\nu$ from $P/R$ to $P$ and $\nu_i$ from $A_i/R_i$ to $A_i$ (for each index $i$) such that $\eta \circ \nu = I_{P/R}$ and $\eta_i \circ \nu_i = I_{A_i/R_i}$ (for each index $i$). Define a mapping $F$ from $P/R$ to $P' = \prod_{i \in I} A_i/R_i$ as follows: for each element $X$ of $P/R$ and each index $i$ in $I$ set

$$(F(X))(i) = \eta_i((\nu(X))(i)).$$

To show that $F$ is injective let $X_1$ and $X_2$ be elements of $P/R$ such that $F(X_1) = F(X_2)$. Then for each index $i$ in $I$ we have

$$\eta_i((\nu(X_1))(i)) = \eta_i((\nu(X_2))(i));$$

hence for every index $i$ in $I$ we have $((\nu(X_1))(i), (\nu(X_2))(i)) \in R_i$ and so, by the definition of $R$, we have $(\nu(X_1), \nu(X_2)) \in R$. It follows that $\eta(\nu(X_1)) = \eta(\nu(X_2))$, i.e. $X_1 = X_2$. So $F$ is injective.

To prove that $F$ is surjective let $f'$ be any element of $P' = \prod_{i \in I} A_i/R_i$. Let $f$ be the element of $P = \prod_{i \in I} A_i$ given by setting

$$f(i) = \nu_i(f'(i)) \text{ for all indices } i \text{ in } I.$$

Let $X = \eta(f)$; we claim that $F(X) = f'$. To see this we notice that $\eta(\nu(\eta(f))) = \eta(f)$ and hence $(\nu(\eta(f)), f) \in R$, i.e. $(\nu(X), f) \in R$. It follows from the definition of $R$ that we have $((\nu(X))(i), f(i)) \in R_i$ for each index $i$ in $I$. So for each index $i$ in $I$ we have

$$(F(X))(i) = \eta_i((\nu(X)))(i) = \eta_i(f(i)) = \eta_i(\nu_i(f'(i))) = f'(i).$$

Thus $F(X) = f'$ and hence $F$ is surjective.

60. Let $(a, b) \in R_f$. Then $f(a) = f(b)$. Now $(a, f(a)) \in F$ and $(b, f(b)) \in F$. So $(a, f(a)) \in F$ and $(f(a), b) = (f(b), b) \in F^{-1}$. Thus $(a, b) \in F^{-1} \circ F$.

    Conversely, suppose $(a, b) \in F^{-1} \circ F$. Then there is an element $t$ of $B$ such that $(a, t) \in F$ and $(t, b) \in F^{-1}$, so $(b, t) \in F$. Then $f(a) = t = f(b)$ and so $(a, b) \in R_f$.

61. $(a, b) \in R$
    $\iff R(a) = R(b)$ (by Exercise 58)
    $\iff \eta(a) = \eta(b)$
    $\iff (a, b) \in$ the equivalence relation associated with $\eta$.

62. (1) Let $A$ be a subset of $E$ such that $\eta^{\leftarrow}(\eta^{\rightarrow}(A)) = A$.

    Let $a$ be any element of $A$; let $b$ be any element of the $R$-class of $a$. Then $\eta(b) = \eta(a)$ and so $b \in \eta^{\leftarrow}(\eta^{\rightarrow}(A)) = A$.

    Thus $A$ is saturated.

    (2) Conversely, suppose $A$ is saturated.

    For every subset $A$ of $E$ we have $\eta^{\leftarrow}(\eta^{\rightarrow}(A)) \supseteq A$. So suppose $a$ is any element of $\eta^{\leftarrow}(\eta^{\rightarrow}(A))$. Then $\eta(a) \in \eta^{\rightarrow}(A)$ and so there is an element $a_1$ of $A$ such that $\eta(a) = \eta(a_1)$. So $a \in R(a_1)$ and hence, since $A$ is saturated, we have $a \in A$. Thus $\eta^{\leftarrow}(\eta^{\rightarrow}(A)) \subseteq A$.

    So finally $\eta^{\leftarrow}(\eta^{\rightarrow}(A)) = A$ as asserted.

63. (1) Suppose there is a mapping $h$ from $E/R$ to $F$ such that $f = h \circ \eta$. Let $(x, y) \in R$. Then $\eta(x) = \eta(y)$ and so $h(\eta(x)) = h(\eta(y))$, whence $f(x) = f(y)$.

    (2) Conversely, suppose that for all $(x, y)$ in $R$ we have $f(x) = f(y)$. Let $\nu$ be a mapping from $E/R$ to $E$ such that $\eta \circ \nu = I_{E/R}$. Let $h = f \circ \nu$. Then $h \circ \eta = f \circ \nu \circ \eta$. For each element $x$ of $E$ we have $\eta(\nu(\eta(x))) = \eta(x)$; so $(\nu(\eta(x)), x) \in R$ and hence $f(\nu(\eta(x))) = f(x)$ and consequently $h \circ \eta = f \circ \nu \circ \eta = f$.

64. (1) Suppose there is a mapping $h$ from $E/R$ to $F/S$ such that $h \circ \eta_R = \eta_S \circ f$. Let $(x, y) \in R$. Then $\eta_S(f(x)) = h(\eta_R(x)) = h(\eta_R(y)) = \eta_S(f(y))$. So $(f(x), f(y)) \in S$.

    (2) Conversely, suppose that for all elements $x$, $y$ of $E$ such that $(x, y) \in R$ we have $(f(x), f(y)) \in S$. Then whenever $(x, y) \in R$ we have $\eta_S(f(x)) = \eta_S(f(y))$, i.e. $(\eta_S \circ f)(x) = (\eta_S \circ f)(y)$. It follows from the second part of Exercise 63 that there is a mapping $h$ from $E/R$ to $F/S$ such that $h \circ \eta_R = \eta_S \circ f$.

65. Let $\eta_R$ and $\eta_S$ be the canonical surjections from $E$ onto $E/R$ and $E/S$ respectively. Let $\nu_R$ and $\nu_S$ be mappings from $E/R$ and $E/S$ respectively to $E$ such that $\eta_R \circ \nu_R = I_{E/R}$ and $\eta_S \circ \nu_S = I_{E/S}$. Define a relation $R/S$ on $E/S$ by setting $(X, Y) \in R/S$ if and only if $(\nu_S(X), \nu_S(Y)) \in R$. Then $R/S$ is an equivalence relation on $E/S$. Let $\eta_{R/S}$ be the canonical surjection from $E/S$ onto $(E/S)/(R/S)$; let $\nu_{R/S}$ be a mapping from $(E/S)/(R/S)$ to $E/S$ such that $\eta_{R/S} \circ \nu_{R/S} = I_{(E/S)/(R/S)}$. Let $f$ be the mapping $\eta_{R/S} \circ \eta_S \circ \nu_R$ from $E/R$ to $(E/R)/(R/S)$.

    To show that $f$ is injective, let $x$ and $y$ be elements of $E/R$ such that $f(x) = f(y)$. Then we have $(\eta_S \circ \nu_R(x), \eta_S \circ \nu_R(y)) \in R/S$; so $(\nu_S \circ \eta_S \circ \nu_R(x), \nu_S \circ \eta_S \circ \nu_R(y)) \in R$. Now $\eta_S \circ \nu_S \circ \eta_S \circ \nu_R(x) = \eta_S \circ \nu_R(x)$; so $(\nu_R(x), \nu_S \circ \eta_S \circ \nu_R(x)) \in S \subseteq R$. Similarly we deduce that $(\nu_S \circ \eta_S \circ \nu_R(y), \nu_R(y)) \in R$. Since $R$ is transitive we deduce that $(\nu_R(x), \nu_R(y)) \in R$ and hence that $\eta_R(\nu_R(x)) = \eta_R(\nu_R(y))$, i.e. $x = y$.

    To show that $f$ is surjective, let $C$ be any element of $(E/S)/(R/S)$. Let $x = \eta_R \circ \nu_S \circ \nu_{R/S}(C)$. Since $\eta_R \circ \nu_R \circ \eta_R \circ \nu_S \circ \nu_{R/S}(C) = \eta_R \circ \nu_S \circ \nu_{R/S}(C)$ it follows that $(\nu_R(x), \nu_S \circ \nu_{R/S}(C)) \in R$. Since $\eta_S \circ \nu_S \circ \eta_S \circ \nu_R(x) = \eta_S \circ \nu_R(x)$ we have $(\nu_S \circ \eta_S \circ \nu_R(x), \nu_R(x)) \in S \subseteq R$. It follows, since $R$ is transitive, that we have $(\nu_S \circ \eta_S \circ \nu_R(x), \nu_S \circ \nu_{R/S}(C)) \in R$. Hence, according to the definition of $R/S$, we have $(\eta_S \circ \nu_R(x), \nu_{R/S}(C)) \in R/S$ and hence

$$f(x) = \eta_{R/S} \circ \eta_S \circ \nu_R(x) = \eta_{R/S} \circ \nu_{R/S}(C) = C.$$

# Chapter 19

# ANSWERS TO CHAPTER 6

66. Suppose $R$ is an order. Then $D_X \subseteq R$, $R \cap R^{-1} \subseteq D_X$ and $R \circ R \subseteq R$.

    Hence $D_X = (D_X)^{-1} \subseteq R^{-1}$, $R^{-1} \cap (R^{-1})^{-1} = R \cap R^{-1} \subseteq D_X$ and $R^{-1} \circ R^{-1} = (R \circ R)^{-1} \subseteq R^{-1}$.

    So $R^{-1}$ is an order on $X$.

67. (1) Suppose $R$ is an order on $X$.

    Then $R \circ R \subseteq R$ and $R \cap R^{-1} \subseteq D_X$. Let $(a, b) \in R$. Since $(b, b) \in R$ we have $(a, b) \in R \circ R$. So $R \subseteq R \circ R$ and hence $R \circ R = R$. Let $(a, a) \in D_X$. Then $(a, a) \in R$ and $(a, a) \in R^{-1}$. So $D_X \subseteq R \cap R^{-1}$ and hence $R \cap R^{-1} = D_X$.

    (2) Conversely, suppose $R \circ R = R$ and $R \cap R^{-1} = D_X$.

    Since $D_X = R \cap R^{-1} \subseteq R$, $R$ is reflexive. Since $R \cap R^{-1} = D_X \subseteq D_X$, $R$ is antisymmetric. Since $R \circ R = R \subseteq R$, $R$ is transitive.

    Thus $R$ is an order.

68. Let $a$ be any element of $A$. Since $A \subseteq B$, we have $R$-inf $B \leq a$. So $R$-inf $B$ is a lower bound for $A$; so, since $R$-inf $A$ is the greatest lower bound for $A$, we have $R$-inf $B \leq R$-inf $A$. Similarly $R$-sup $A \leq R$-sup $B$.

69. Suppose $A$ has an $R$-least upper bound, $a$.

    Then for each index $i$ in $I$ we have that $a$ is an $R$-upper bound for $A_i$ and hence $b_i \leq a$. Thus $a$ is an R-upper bound for $B$. Let

$c$ be any $R$-upper bound for $B$. If $x$ is any element of $A$ there is an index $i$ in $I$ such that $x \in A_i$. Then $x \leq b_i \leq c$. So $c$ is an $R$-upper bound for $A$, and so $a \leq c$.

Hence $a$ is $R$-least upper bound for $B$.

Conversely, suppose $B$ has an $R$-least upper bound, $b$.

For each element $x$ of $A$ there is an index $i$ in $I$ such that $x \in A_i$. Then $x \leq b_i \leq b$. So $b$ is an $R$-upper bound for $A$. Let $d$ be any $R$-upper bound for $A$. Then $d$ is an $R$-upper bound for each of the subsets $A_i$ and so $d \geq b_i$ for all indices $i$. Hence $d$ is an $R$-upper bound for $B$. Thus $d \geq b$.

So $b$ is the $R$-least upper bound for $A$.

70. Let $S$ and $T$ be subspaces of $V$.

    Then $S \cap T$ is the $(\subseteq)$-greatest lower bound of $\{S, T\}$ and the subspace $S + T$ spanned by $S \cup T$ is the $(\subseteq)$-least upper bound of $\{S, T\}$.

71. For each pair of indices $i_0$ in $I$, $j_0$ in $J$ we have

$$\inf_{i \in I} x_{ij_0} \leq x_{i_0 j_0} \leq \sup_{j \in J} x_{i_0 j}.$$

Thus for each index $j_0$ in $J$ the element $\inf_{i \in I} x_{ij_0}$ is a lower bound of the set $\{\sup_{j \in J} x_{ij}\}_{i \in I}$. Hence for each $j_0$ in $J$ we have $\inf_{i \in I} x_{ij_0} \leq \inf_{i \in I}(\sup_{j \in J} x_{ij})$. Thus $\inf_{i \in I}(\sup_{j \in J} x_{ij})$ is an upper bound for the set $\{\inf_{i \in I} x_{ij}\}_{j \in J}$.

Hence $\sup_{j \in J}(\inf_{i \in I} x_{ij}) \leq \inf_{i \in I}(\sup_{j \in J} x_{ij})$.

72. (1) Let $T$ be a total order on $E$.

    Let $S$ be any order on $E$ such that $S \supseteq T$. If $(x, y) \in S$, then either $(x, y) \in T$ or $(y, x) \in T$. If $(y, x) \in T$ then we have $(y, x) \in S$ and $(x, y) \in S$; so $x = y$ and hence $(x, y) \in T$.

    Thus $S = T$, i.e. $T$ is $(\subseteq)$-maximal in $Q$.

    (2) Conversely, suppose that $T$ is an order on $E$ but not a total order. Thus there are elements $a$, $b$ of $E$ such that $(a, b) \notin T$ and $(b, a) \notin T$. Let

$$T' = T \cup \{(x, y) : ((x, y) \in E \times E) \wedge ((x, a) \in T) \wedge ((b, y) \in T)\}.$$

Then $(a, b) \in T'$ but $(a, b) \notin T$; so $T' \supset T$. We shall prove that $T'$ is an order on $E$; this shows that $T$ is not $(\subseteq)$-maximal in $Q$.

(a) $T' \supset T \supseteq D_E$; so $T'$ is reflexive.

(b) Suppose $(x, y) \in T'$ and $(y, x) \in T'$.

If $(x, y) \in T$ and $(y, x) \in T$ then $x = y$ since $T$ is an order.

If $(x, y) \in T$ and $(y, x) \notin T$ then $(x, y) \in T$, $(y, a) \in T$ and $(b, x) \in T$. Since $T$ is transitive this yields $(b, a) \in T$, which is a contradiction. So this case cannot occur.

If $(x, y) \notin T$ and $(y, x) \in T$ then $(x, a) \in T$, $(b, y) \in T$ and $(y, x) \in T$. Again we have $(b, a) \in T$; so this case cannot occur.

Finally, if $(x, y) \notin T$ and $(y, x) \notin T$ then $(x, a) \in T$, $(b, y) \in T$, $(y, a) \in T$ and $(b, x) \in T$. Again we have $(b, a) \in T$; so this case cannot occur.

Hence $T'$ is antisymmetric.

(c) Suppose $(x, y) \in T'$ and $(y, z) \in T'$.

If $(x, y) \in T$ and $(y, z) \in T$ then $(x, z) \in T$ and hence $(x, z) \in T'$.

If $(x, y) \in T$ and $(y, z) \notin T$ then $(x, y) \in T$, $(y, a) \in T$ and $(b, z) \in T$. We have then $(x, a) \in T$ and $(b, z) \in T$; so $(x, z) \in T'$.

If $(x, y) \notin T$ and $(y, z) \in T$ then $(x, a) \in T$, $(b, y) \in T$ and $(y, z) \in T$. We deduce that $(x, a) \in T$ and $(b, z) \in T$; so $(x, z) \in T'$.

If $(x, y) \notin T$ and $(y, z) \notin T$ then we have $(x, a) \in T$, $(b, y) \in T$, $(y, a) \in T$ and $(b, z) \in T$. From these we deduce that $(b, a) \in T$, which is not the case; hence this possibility cannot occur.

It follows that $T'$ is transitive.

So $T'$ is an order and hence $T$ is not $(\subseteq)$-maximal.

73. Let $x \in E$. Then $(x, x)$ belongs to every order in $S$ and hence $(x, x) \in \bigcup S$. Thus $\bigcup S$ is reflexive.

Suppose $(x, y) \in \bigcup S$ and $(y, x) \in \bigcup S$. Then there are orders $R_1$ and $R_2$ in $S$ such that $(x, y) \in R_1$ and $(y, x) \in R_2$. Since $S$ is totally ordered by inclusion we have either $R_1 \subseteq R_2$ or $R_2 \subseteq R_1$. If $R_1 \subseteq R_2$ then $(x, y) \in R_2$ and $(y, x) \in R_2$; since $R_2$ is an order we have $x = y$. A similar argument holds if $R_2 \subseteq R_1$.

Thus $\bigcup S$ is antisymmetric.

Suppose $(x, y) \in \bigcup S$ and $(y, z) \in \bigcup S$. Then there are orders $R_1$ and $R_2$ in $S$ such that $(x, y) \in R_1$ and $(y, z) \in R_2$. As above, we have either $R_1 \subseteq R_2$ or $R_2 \subseteq R_1$. If $R_1 \subseteq R_2$ then we have $(x, y) \in R_2$ and $(y, z) \in R_2$; hence, since $R_2$ is an order we have $(x, z) \in R_2$ and hence $(x, z) \in \bigcup S$. A similar argument holds if $R_2 \subseteq R_1$.

Thus $\bigcup S$ is transitive.

74. Let $x$ be any element of $E$.

    Then $f(x) \in E'$ and so $(f(x), f(g(f(x)))) \in R'$. Since we have $(x, g(f(x))) \in R$ and $f$ is decreasing, it follows that $(f(g(f(x))), f(x)) \in R'$. Hence $f(g(f(x))) = f(x)$ for all $x$ in $E$ and so $f \circ g \circ f = f$.

    Similarly $g \circ f \circ g = g$.

75. (1) Suppose $f$ is increasing. Let $x$ and $y$ be any elements of $E$. Since $(R\text{-inf } \{x, y\}, x) \in R$ and $(R\text{-inf } \{x, y\}, y) \in R$ we have $(f(R\text{-inf } \{x, y\}), f(x)) \in R'$ and $(f(R\text{-inf } \{x, y\}), f(y)) \in R'$. Thus $f(R\text{-inf } \{x, y\})$ is an $R'$-lower bound for $\{f(x), f(y)\}$. Hence we have $(f(R\text{-inf } \{x, y\}), R'\text{-inf } \{f(x), f(y)\}) \in R'$.

    (2) Conversely, suppose the condition holds. If $(x, y) \in R$ then $R\text{-inf } \{x, y\} = x$. So $(f(x), R'\text{-inf } \{f(x), f(y)\}) \in R'$. Hence $(f(x), f(y)) \in R'$.

    Thus $f$ is increasing.

# Chapter 20

# ANSWERS TO CHAPTER 7

76. Let $R$ be a well-ordering on a class $X$.

    Let $a$ and $b$ be elements of $X$. Then the subset $\{a, b\}$ has an $R$-least member. If the $R$-least member is $a$ then $(a, b) \in R$; if the $R$-least member is $b$ then $(b, a) \in R$.

    Thus $R$ is a total order on $X$.

77. (1) Let $a \in A$. Then $(a, a) \in R \cap (A \times A) = R|A$.

    Let $a$ and $b$ be elements of $A$ such that we have $(a, b) \in R|A$ and $(b, a) \in R|A$. Then $(a, b) \in R$ and $(b, a) \in R$. Hence $a = b$.

    Let $a$, $b$, $c$ be elements of $A$ for which $(a, b) \in R|A$ and $(b, c) \in R|A$. Then $(a, b) \in R$ and $(b, c) \in R$. So $(a, c) \in R$. But of course $(a, c) \in A \times A$. So $(a, c) \in R|A$.

    Thus $R|A$ is an order on $A$.

    (2) Suppose $R$ is a total order on $A$.

    Let $a$ and $b$ be any elements of $A$. Then $(a, b) \in A \times A$. Since $R$ is a total order on $X$ we have either $(a, b) \in R$ or $(b, a) \in R$. Hence either $(a, b) \in R|A$ or $(b, a) \in R|A$.

    Hence $R|A$ is a total order on $A$.

    (3) Suppose $R$ is a well-ordering on $X$.

    Let $B$ be a non-empty subset of $A$. Then $B$ has an $R$-least member, $b_0$ say, i.e. $(b_0, b) \in R$ for all $b$ in $B$. So $(b_0, b) \in R|A$ for all $b$ in $B$, i.e. $b_0$ is $(R|A)$-least member of $B$.

78. For each element $x$ of $X$ we have $x \in X$ but $x \notin S_{X,R}(x)$. So $S_{X,R}(x) \subset X$.

    (1) Let $t$ be any member of $S_{A,R|A}(a)$. Then we have $t \in A$, $t \neq a$ and $(t, a) \in R\,|A$.

    It follows that $t \in X$, $t \neq a$ and $(t, a) \in R$, i.e. $t \in S_{X,R}(a)$. So $S_{A,R|A}(a) \subseteq S_{X,R}(a)$.

    (2) Let $t$ be any element of $S_{X,R}(a)$. Then $t \neq a$ and $(t, a) \in R$. Since $(a, b) \in R$ it follows that $(t, b) \in R$ and $t \neq b$, i.e. that $t \in S_{X,R}(b)$. So $S_{X,R}(a) \subseteq S_{X,R}(b)$.

    (3) As in (1), $S_{S,R|S}(x) \subseteq S_{X,R}(x)$.

    Conversely, let $t$ be any element of $S_{X,R}(x)$. Then $t \neq x$ and $(t, x) \in R$.

    Since $x \in S = S_{X,R}(a)$, which is an $R$-segment, it follows that $t \in S$. Thus $(t, x) \in R \cap (S \times S) = R\,|S$, and so $t \in S_{S,R|S}(x)$. So $S_{X,R}(x) \subseteq S_{S,R|S}(x)$.

79. Let $S$ be a proper $R$-segment of $X$.

    Then $X \sim S$ is non-empty and hence (since $R$ is a well-ordering) has an $R$-least member, say $a$. We claim that $S = S_{X,R}(a)$.

    If $x \in S_{X,R}(a)$ then $x < a$ and hence $x \notin X \sim S$ (for if $x$ were in $X \sim S$ we would have $a \leq x$). So $x \in S$. Thus $S_{X,R}(a) \subseteq S$.

    Conversely, if $x \notin S_{X,R}(a)$ then $x \nless a$ and so $x \geq a$. It follows that $x \notin S$ (for if $x$ were in $S$ then, since $S$ is a segment, we would have $a$ in $S$). Thus $S \subseteq S_{X,R}(a)$.

    So $S = S_{X,R}(a)$ as asserted.

80. Let $A$ be the set of all $R$-segments of $E$. Let $A^\star$ be the set of all proper $R$-segments. It will be sufficient to prove that $A^\star$ is well-ordered by the inclusion relation.

    Let $B$ be any non-empty subset of $A^\star$.

    Let $T = \{x : S_{E,R}(x) \in B\}$. Then (since every $R$-segment in $A^\star$ is an initial segment) $T$ is non-empty and hence has an $R$-least element, say $t_0$. We claim that $S_{E,R}(t_0)$ is the $(\subseteq)$-least member of $T$. For if $S_{E,R}(x) \in B$ we have $x \in T$ and so $t_0 \leq x$, whence $S_{E,R}(t_0) \subseteq S_{E,R}(x)$.

81. Suppose, to the contrary, that $A \neq X$.

Then $X \sim A$ is non-empty and hence has an $R$-least element, $b$ say. Then $S_{X,R}(b) \subseteq A$. (For if $x \in S_{X,R}(b)$ we have $x < b$, while if $x \notin A$ we have $x \geq b$.) By hypothesis it follows that $b \in A$, which is a contradiction.

So $A = X$ as asserted.

82. Suppose, to the contrary, that $\{t : (t \in E) \wedge ((F(t), G(t)) \notin R')\}$ is non-empty.

Then there is an $R$-least element of this class, say $a$. For every element $x$ of $E$ such that $x <_R a$ we have

$$F(x) \leq_{R'} G(x) <_{R'} G(a) <_{R'} F(a) \in F^{\rightarrow}(E)$$

(by condition (2)). Since $F^{\rightarrow}(E)$ is a segment it follows that $G(a) \in F^{\rightarrow}(E)$; so there is an element $y$ of $E$ such that $G(a) = F(y)$. Since $G(a) = F(y) <_{R'} F(a)$ we must have $y <_R a$. It follows that $G(a) = F(y) \leq_{R'} G(y) <_{R'} G(a)$, which is a contradiction.

So $F(x) \leq_{R'} G(x)$ for all $x$ in $E$.

83. Apply Theorem 4 to the graphs of the mappings $f$ and $g$.

84. Let $f$ and $g$ be isomorphisms from $E$ onto $E'$. Then $f$ and $g$ are strictly increasing and their ranges $(= E')$ are segments of $E'$. Thus (by Theorem 4) $f(x) < g(x)$ and $g(x) < f(x)$ for all $x$ in $E$. So $f(x) = g(x)$ for all $x$ in $E$, i.e. $f = g$.

85. Suppose there were an isomorphism $f$ from $E$ onto an initial segment $S = S_{E,R}(a)$ of $E$; $f$ is strictly increasing. The identity mapping $I_E$ from $E$ to $E$ is increasing and its range $(= E)$ is a segment. Thus for every element $x$ of $E$ we have $I_E(x) = x \leq f(x)$. In particular we have $a \leq f(a) < a$, which is a contradiction.

86. The uniqueness parts of the statement follow from Exercise 84.

Let $A = \{x : (x \in E)$ and there exists an isomorphism from $S_{E,R}(x)$ onto an initial segment of $E'\}$. If $x \in A$ there is a unique element $x'$ in $E'$ such that there is an isomorphism from $S_{E,R}(x)$ onto $S_{E',R'}(x')$. Define a mapping $f$ from $A$ to $E'$ by setting

$f(x) =$ the corresponding $x'$ for each $x$ in $E$. Let $A' = f^{\rightarrow}(A)$; clearly $f$ produces an isomorphism from $A$ onto $A'$.

There are four possibilities to consider: (1) $A = E$ and $A' = E'$, (2) $A = E$ and $A' \subset E'$, (3) $A \subset E$ and $A' = E'$, (4) $A \subset E$ and $A' \subset E'$.

In Case (1), $f$ is an isomorphism from $E$ onto $E'$.

Now consider Case (2). We claim that $A'$ is an initial segment of $E'$. Certainly, since $A' \neq E'$, it follows that $E' \sim A'$ is non-empty and hence has a least element, $b'$ say. Then $S_{E',R'}(b') \subseteq A'$; for if $x' \in S_{E',R'}(b')$ we have $x' < b'$; hence $x' \notin E' \sim A'$ and so $x' \in A'$.

Conversely, suppose $x' \in A'$. If $x' \notin S_{E',R'}(b')$ then $x' \geq b'$. Since $x' \in A'$ there is an isomorphism $\varphi$ from $S_{E,R}(f^{-1}(x'))$ onto $S_{E',R'}(x')$. Let $b = f^{-1}(b')$; then it is clear that the restriction of $\varphi$ to $S_{E,R}(b)$ is an isomorphism from $S_{E,R}(b)$ onto $S_{E',R'}(b')$. Thus $b' \in f^{\rightarrow}(A) = A'$, which is a contradiction.

Thus in Case (2) there is an isomorphism from $E$ onto the initial segment $S_{E',R'}(b')$ of $E'$.

Similarly in Case (3) there is an isomorphism from an initial segment of $E$ onto $E'$, the inverse of which is an isomorphism from $E'$ onto an initial segment of $E$.

Finally we show that Case (4) cannot occur. If we have $A \neq E$ and $A' \neq E'$ then (as in Case (2)) $A = S_{E,R}(b)$ and $A' = S_{E',R'}(b')$ where $b$ and $b'$ are the least elements of $E \sim A$ and $E' \sim A'$ respectively. Since $f$ is an isomorphism from $A$ to $A'$ we have $b \in A$ and $b' \in A'$ which is a contradiction.

# Chapter 21

# ANSWERS TO CHAPTER 8

87. Routine checking.

88. Let $A = \{x : (x \in X) \wedge (f(x) = x)\}$. We shall show that $A = X$.

    So suppose, to the contrary, that $A \neq X$ and let $u$ be the least element of $X \sim A$. Then for all elements $x$ of $X$ such that $x \in u$ we have $x \in A$ and so $f(x) = x$. We claim that $u = f(u)$.

    Let $x \in u$; then, as we remarked above, we have $f(x) = x$. But since $f$ is an isomorphism we have $f(x) \in f(u)$; so $x \in f(u)$. Thus $u \subseteq f(u)$.

    Conversely, let $x \in f(u)$. Then $x \in Y$ ($f(u)$ is an element of $Y$, hence a subset of $Y$). So there is an element $v$ of $X$ such that $x = f(v)$. Since $f$ is an isomorphism, $f(v) \in f(u)$ implies that $v \in u$ and hence $v = f(v) = x$; so $x \in u$. Thus $f(u) \subseteq u$.

    Thus $f(u) = u$ and so $u \in A$, which is a contradiction.

89. Since $x$ is an element of the ordinal $X$, $x$ is a subset of $X$ (in particular, a set). Since $E\,|X$ is a well-ordering on $X$, we see that $E\,|x$ is a well-ordering on $x$.

    Let $u$ be any element of $x$. Since $x \subseteq X$ we have $u \in X$ and hence $u \subseteq X$. To show that $u \subseteq x$ let $v$ be any element of $u$. Since $u \subseteq X$ we have $v \in X$. Now $u$, $v$, $x$ are elements of $X$ and we have $u \in x$, $v \in u$, i.e. $(u, x) \in E\,|X$ and $(v, u) \in E\,|X$. It follows that $(v, x) \in E\,|X$ and so $v \in x$. Thus $u \subseteq x$.

    So $x$ is an ordinal number.

To show that $x = S_{X,E|X}(x)$ notice first that if $y \in S_{X,E|X}(x)$ then $(y, x) \in E|X$ and $y \neq x$; so $y \in x$; secondly, if $y \in x$ then $y \in X$ (since $x \subset X$) and so $y \in S_{X,E|X}(x)$.

90. (1) Let $\alpha$ be any element of $On$, i.e. any ordinal number. Then $\alpha = \alpha$, so $(\alpha, \alpha) \in E|On$; hence $E|On$ is reflexive.

Let $\alpha$ and $\beta$ be ordinal numbers such that $(\alpha, \beta) \in E|On$ and $(\beta, \alpha) \in E|On$. If $\alpha \neq \beta$ we have $\beta \in \alpha$ and $\alpha \in \beta$. Thus $\alpha \subseteq \beta$ and so $\beta \in \beta$, which contradicts the Axiom of Foundation. Hence $\alpha = \beta$ and thus $E|On$ is antisymmetric.

Let $\alpha$, $\beta$, $\gamma$ be ordinal numbers such that $(\alpha, \beta) \in E|On$ and $(\beta, \gamma) \in E|On$. If $\alpha = \beta$ or $\beta = \gamma$ we have $(\alpha, \gamma) \in E|On$. Suppose now that $\alpha \in \beta$ and $\beta \in \gamma$. Since $\gamma$ is an ordinal we have $\beta \subseteq \gamma$ and hence $\alpha \in \gamma$, so $(\alpha, \gamma) \in E|On$. So $E|On$ is transitive.

To show that $On$ is well-ordered by $E|On$, let $x$ be any non-empty subset of $On$. Let $\alpha$ be any ordinal number in $x$. If $\alpha$ is the $(E|On)$-least element of $x$ we are done; if not, there is an element $\beta$ of $x$ such that $\beta \in \alpha$. Thus $\alpha \cap x \neq \emptyset$. Now $E|\alpha$ is a well-ordering on $\alpha$ and hence $\alpha \cap x$ has an $(E|\alpha)$-least element, $\gamma$ say. We claim that $\gamma$ is the $(E|On)$-least element of $x$. For suppose there were an element $\delta$ of $x$ such that $\delta \in \gamma$. Since $\gamma \in \alpha$ we have $\gamma \subseteq \alpha$ and hence $\delta \in \alpha$. Thus $\delta \in \alpha \cap x$ and $\delta \in \gamma$, contradicting the definition of $\gamma$ as $(E|\alpha)$-least element of $\alpha \cap x$. So $\gamma$ is $(E|On)$-least element of $x$ and hence $E|On$ is a relation of well-ordering on $On$.

(2) Let $\alpha$ be any element of $On$, i.e. any ordinal number.

If $x$ is any element of $\alpha$ then (by Exercise 89) $x$ is an ordinal number, i.e. $x \in On$. So $\alpha \subseteq On$.

Thus $On$ is an ordinal.

91. If $On$ were a set, then (according to the previous Exercise) it would be an ordinal number and so we would have $On \in On$, contradicting the Axiom of Foundation.

92. This follows as a special case of Exercise 89.

93. If $x$ is any element of $\alpha$ then, since $\alpha \in \alpha^+$ and $x \in \alpha$ we have $x \in \bigcup \alpha^+$. So $\alpha \subseteq \bigcup \alpha^+$.

Conversely, if $x$ is any element of $\bigcup \alpha^+$ there is an element $y$ of $\alpha^+$ such that $x \in y$. So either (1) $x \in \alpha$ or (2) $x \in y$ where $y \in \alpha$. In case (2), since $\alpha$ is an ordinal number and $y \in \alpha$, we have $y \subseteq \alpha$; hence $x \in \alpha$. Thus $\bigcup \alpha^+ \subseteq \alpha$.

94. Since $\beta$ is a set we deduce (using the Pairing and Union Axioms) that $\beta^+ = \beta \cup \{\beta\}$ is a set. $E\,|\beta^+$ is easily seen to be a well-ordering on $\beta^+$. (If $x$ is any non-empty subset of $\beta^+$ then (1) if $x \cap \beta \neq \emptyset$ the $(E\,|\beta^+)$-least element of $x$ is the $(E\,|\beta)$-least element of $x \cap \beta$ and (2) if $x \cap \beta = \emptyset$ the $(E\,|\beta^+)$-least element of $x$ is $\beta$.)

    If $x \in \beta$ then, since $\beta$ is an ordinal, we have $x \subseteq \beta$ and so $x \subseteq \beta^+$. If $x = \beta$ then we have $x \subseteq \beta^+ = \beta \cup \{\beta\}$. Hence every element of $\beta^+$ is a subset of $\beta^+$.

    Thus $\beta^+$ is an ordinal number.

95. (1) Suppose $\alpha$ is a limit ordinal.

    Let $\beta$ be any element of $\alpha$. Since $E\,|On$ is a well-ordering on $On$ we have either (1) $\beta^+ \in \alpha$ or (2) $\beta^+ = \alpha$ or (3) $\alpha \in \beta^+$.

    Case (2) is ruled out since $\alpha$ is a limit ordinal. If $\alpha \in \beta^+$ then either $\alpha \in \beta$ (which is ruled out since $\beta \in \alpha$) or $\alpha = \beta$ (which is ruled out since we do not have $\beta \in \beta$). Thus Case (3) is ruled out.

    Hence only Case (1) remains and so $\beta^+ \in \alpha$.

    (2) Conversely, suppose that for every element $\beta$ of $\alpha$ we have $\beta^+ \in \alpha$. If $\alpha$ were a non-limit ordinal, say $\alpha = \gamma^+ = \gamma \cup \{\gamma\}$, we would have $\gamma \in \alpha$ but $\gamma^+ \notin \alpha$ contradicting our hypothesis.

    Hence $\alpha$ is a limit ordinal.

96. Applying the extension of Exercise 86 described in Chapter 8 to the well-ordered set $(x, R)$ and the class $On$ well-ordered by $E$ we see that there are three possibilities—either (1) there is a unique isomorphism from $x$ onto $On$ or (2) there is a unique isomorphism from $x$ onto an initial segment of $On$ or (3) there is a unique isomorphism from $On$ onto an initial segment of $x$.

    (1) and (3) are ruled out, since $x$ and its initial segments are sets, while $On$ is a proper class. (The Replacement Axiom assures us that if the domain of a functional relation is a set then so is its range.) Hence there is a unique $\alpha$ in $On$ and an isomorphism from $x$ onto $S_{On,E\,|On}(\alpha) = \alpha$.

97. Let $\alpha = \text{Ord}(x, R)$ and $\beta = \text{Ord}(y, S)$.

Then there exists an $(R, E \mid \alpha)$-isomorphism $f$ from $x$ to $\alpha$ and an $(S, E \mid \beta)$-isomorphism from $y$ to $\beta$.

If $\alpha = \beta$ then $g^{-1} \circ f$ is an $(R, S)$-isomorphism from $x$ to $y$.

Conversely, if there is an $(R, S)$-isomorphism $h$ from $x$ to $y$ then $g \circ h \circ f^{-1}$ is an $(E \mid \alpha, E \mid \beta)$-isomorphism from $\alpha$ to $\beta$. By Exercise 88 it follows that $\alpha = \beta$.

98. (1) Suppose $R$ is a relation of well-ordering on $x$. According to Theorem 4 there is a unique ordinal number $\alpha$ such that $(\alpha, E|\alpha)$ is isomorphic to $(x, R)$. Then of course $x$ is equinumerous with the ordinal number $\alpha$.

(2) Conversely, suppose $x$ is equinumerous with an ordinal number $\alpha$. Let $f$ be a bijection from $x$ onto $\alpha$. Then the relation $R$ on $x$ defined by setting $(a, b) \in R$ if and only if $(f(a), f(b)) \in E$ is clearly a well-ordering on $x$.

99. For each ordinal number $\alpha$ we shall write $S(\alpha)$ as an abbreviation for $S_{On,E|On}(\alpha)$.

Let $M$ be the class of all functional relations $H$ such that

(1) Dom $H \subseteq On$;

(2) for all ordinal numbers $\alpha$ in Dom $H$ we have $\alpha \subseteq$ Dom $H$ and $H(\alpha) = F(H^{\rightarrow}(\alpha))$.

We claim that if $H_1$ and $H_2$ are relations in $M$ then $H_1(\alpha) = H_2(\alpha)$ for all ordinal numbers $\alpha$ in Dom $H_1 \cap$ Dom $H_2$. If this is not the case then the class

$$\{\alpha : (\alpha \in \text{Dom } H_1 \cap \text{ Dom } H_2) \wedge (H_1(\alpha) \neq H_2(\alpha))\}$$

is non-empty and hence has an $E$-least member, $\alpha_0$ say. Then, since $\alpha_0 \in$ Dom $H_1$ we have $\alpha_0 \subseteq$ Dom $H_1$ and similarly $\alpha_0 \subseteq$ Dom $H_2$. For each element $\beta$ of $\alpha_0$ we have $\beta$ $E$-less then $\alpha_0$ and hence $H_1(\beta) = H_2(\beta)$. So $H_1^{\rightarrow}(\alpha_0) = H_2^{\rightarrow}(\alpha_0)$ and hence $H_1(\alpha_0) = F(H_1^{\rightarrow}(\alpha_0)) = F(H_2^{\rightarrow}(\alpha_0)) = H_2(\alpha_0)$ which is a contradiction.

Applying the ideas of Exercise 43 to the family of functional relations $(H)_{H \in M}$ we deduce that there is a functional relation $G$ with domain $\bigcup_{H \in M}$ Dom $H$ such that for each $H$ in $M$ and each element $\alpha$ of Dom $H$ we have $G(\alpha) = H(\alpha)$. For every element $\alpha$

of Dom $G$ there is a relation $H$ in $M$ such that $\alpha \in$ Dom $H$, $\alpha \subseteq$ Dom $H$ and $G(\alpha) = H(\alpha) = F(H^{\rightarrow}(\alpha)) = F(G^{\rightarrow}(\alpha))$. Since for every element $\alpha$ of Dom $G$ we have $S(\alpha) = \alpha \subseteq$ Dom $G$ it follows that Dom $G$ is a segment of $On$. We claim that Dom $G = On$. If not, then it is a proper segment and hence, by Exercise 79, we have Dom $G = \gamma$ for some ordinal number $\gamma$. Define the functional relation $G'$ with domain $\gamma^+ = \gamma \cup \{\gamma\}$ consisting of all the ordered pairs $(\alpha, G(\alpha))$ with $\alpha$ in Dom $G$ and the ordered pair $(\gamma, F(G^{\rightarrow}(\gamma)))$. Then $G'$ is in the class $M$ but its domain properly includes Dom $G$ and this is a contradiction since Dom $G = \bigcup_{H \in M}$ Dom $H$.

To establish the uniqueness of $G$ let us suppose that there were two functional relations $G_1$ and $G_2$ with domain $On$ such that $G_1(\alpha) = F(G_1^{\rightarrow}(\alpha))$ and $G_2(\alpha) = F(G_2^{\rightarrow}(\alpha))$ for every ordinal number $\alpha$. If $G_1 \neq G_2$ then the class $\{\alpha : (\alpha \in On) \wedge (G_1(\alpha) \neq G_2(\alpha))\}$ is non-empty and hence has an $E$-least element, $\alpha_o$ say. Then for every ordinal number $\beta$ in $\alpha_0$ we must have $G_1(\beta) = G_2(\beta)$; hence $G_1^{\rightarrow}(\alpha_0) = G_2^{\rightarrow}(\alpha_0)$ and consequently $G_1(\alpha_0) = F(G_1^{\rightarrow}(\alpha_0)) = F(G_2^{\rightarrow}(\alpha_0)) = G_2(\alpha_0)$, which is a contradiction.

# Chapter 22

# ANSWERS TO CHAPTER 9

100. The null class is a member of the set $y$ whose existence is asserted by the Axiom of Infinity; hence $\emptyset$ is a set.

101. Let $n$ be any natural number, $m$ an element of $n$.

     Then $m$ is an ordinal number (by Exercise 89). Further, since $m$ is a subset of $n$ and thus every subset of $m$ is also a subset of $n$, it follows that every non-empty subset of $m$ has an $E$-greatest member.

     Thus $m$ is a natural number.

102. Suppose $\alpha$ is an ordinal with $E$-greatest element $\beta$.

     If $x$ is any element of $\alpha$ then we have $(x, \beta) \in E$ and hence either $x \in \beta$ or $x = \beta$; it follows that $x \in \beta \cup \{\beta\}$.

     Conversely, since $\beta$ is an element of the ordinal $\alpha$ we have $\beta \subseteq \alpha$; since $\beta \in \alpha$ we have $\{\beta\} \subseteq \alpha$. So $\beta \cup \{\beta\} \subseteq \alpha$.

     Hence $\alpha = \beta^+$.

103. Let $n$ be a natural number.

     Then $n$ is an ordinal number and hence $n^+$ is an ordinal number (by Exercise 94). Since $n$ is a natural number every non-empty subset of $n$ has an $E$-greatest element; the $E$-greatest element of $n^+$ is $n$.

     So $n^+$ is a natural number.

104. $\emptyset$ is an ordinal: the defining properties are satisfied vacuously. $\emptyset$ is a set by Exercise 100. So $\emptyset$ is an ordinal number. Since $\emptyset$ has no non-empty subset the second defining property for natural numbers is satisfied vacuously.

     Thus $\emptyset$ is a natural number.

     If $\emptyset$ were the successor of a natural number $n$ we would have $\emptyset = n \cup \{n\}$ and so $n \in \emptyset$, which is a contradiction. So $\emptyset$ is not the successor of any natural number.

105. If $n^+ = m^+$ then we have $\bigcup n^+ = \bigcup m^+$ and hence (by Exercise 93) we have $n = m$.

106. Suppose $x \neq \omega$, so that $\omega \sim x$ is non-empty.

     Let $\alpha$ be the $E$-least element of $\omega \sim x$. Since $\emptyset \in x$ we have $\emptyset \notin \omega \sim x$ and so $\alpha \neq \emptyset$. Since $\alpha$ is a non-empty element, hence subset, of a natural number, it has an $E$-greatest element, $\beta$ say. Then, by Theorem 1, it follows that $\alpha = \beta^+$.

     Since $\beta \in \alpha$ we have $(\beta, \alpha) \in E$ and since $\alpha$ is the $E$-least element of $\omega \sim x$ we must have $\beta \in x$. But then, by condition (2) on the class $x$, we have that $\beta^+ \in x$, i.e. that $\alpha \in x$, which is a contradiction.

     So $x = \omega$, as asserted.

107. By the Axiom of Infinity there is a set $y$ such that $\emptyset \in y$ and whenever we have $x \in y$ then $x^+ \in y$. Then $y \cap \omega$ satisfies the conditions of Theorem 5 and so $y \cap \omega = \omega$. Hence $\omega \subseteq y$ and since $y$ is a set it follows that $\omega$ is a set.

     $E|\omega$ is a relation of well-ordering on $\omega$. If $n \in \omega$ (so that $n$ is a natural number) we have $n \subseteq \omega$ (according to Exercise 101 each element $m$ of the natural number $n$ is a natural number, i.e. $m \in \omega$). So $\omega$ is an ordinal.

     Thus $\omega$ is an ordinal number.

108. Let $x = \{n : (n \in \omega) \wedge P(n)\}$.

     Since $P(\emptyset)$ holds, we have $\emptyset \in x$.

     If $n \in \omega$ and $P(n)$ holds then it follows from (2) that $P(n^+)$ is true; hence $n^+ \in x$.

     Consequently $x = \omega$ and the result follows.

109. It is easy to show that if $g_1$ and $g_2$ are mappings from $\omega$ to $E$ such that $g_1(0) = a = g_2(0)$ and $g_1(n^+) = f(g_1(n))$, $g_2(n^+) = f(g_2(n))$ for all natural numbers $n$ then $g_1 = g_2$. Namely we let $P(n)$ be the open sentence $g_1(n) = g_2(n)$. Then we certainly have $P(0)$ true. Next, let $k$ be any natural number for which $P(k)$ holds, i.e. such that $g_1(k) = g_2(k)$; then $g_1(k^+) = f(g_1(k)) = f(g_2(k)) = g_2(k^+)$ and so $P(k^+)$ is true. It follows from the Principle of Mathematical Induction that $g_1(n) = g_2(n)$ for all natural numbers $n$, i.e. $g_1 = g_2$.

To prove the existence of a mapping satisfying the conditions of Exercise 109 let $M$ be the set of all functional relations $H$ such that Dom $H \subseteq \omega$, $(0, a) \in H$, and for all natural numbers $n$ if $(n, x) \in H$ then $(n^+, f(x)) \in H$.

Let $G = \bigcup M$; we shall show first that $G \in M$.

Clearly Dom $G \subseteq \omega$.

Since $(0, a)$ belongs to every member $H$ of $M$ we have $(0, a) \in G$.

Suppose that $(n, x) \in G$. Then for every member $H$ of $M$ we have $(n, x) \in H$ and hence $(n^+, f(x)) \in H$. Thus $(n^+, f(x)) \in G$.

So $G \in M$ as asserted.

$G$ is clearly a functional relation since all the elements of $M$ are.

We claim that Dom $G = \omega$.

Clearly $0 \in$ Dom $G$.

If $k$ is any natural number in Dom $G$ there is an element $x$ of $E$ such that $(k, x) \in G$. Then $(k^+, f(x)) \in G$ and so $k^+ \in$ Dom $G$.

Applying the Principle of Mathematical Induction to the open sentence $P(n) = (n \in \text{Dom } G)$ we deduce that Dom $G = \omega$ and hence that $g = ((\omega, E), G)$ is a mapping with the properties required.

# Chapter 23

# ANSWERS TO CHAPTER 10

110. Let $(x_i)_{i\in I}$ be a family of non-empty sets.

If $I=\emptyset$ then $\prod_{i\in I} x_i = \mathrm{Map}\,(\emptyset, \bigcup_{i\in I} x_i) \neq \emptyset$.

If $I \neq \emptyset$ let $x = \bigcup_{i\in I} x_i$ and let $R$ be a relation of well-ordering on $x$. Let $s$ be the mapping from $I$ to $x$ given by setting $s(i) = R$-least member of $x_i$ for each $i$ in $I$. Then $s \in \prod_{i\in I} x_i$.

111. Suppose the Axiom of Choice holds.

Let $x$ be any set.

If $x=\emptyset$ then $\emptyset$ is a relation of well-ordering on $x$.

So suppose $x\neq\emptyset$. Let $\mathbf{P}'(x)$ be the set of non-empty subsets of $x$. According to the Axiom of Choice there exists a mapping $f_1$ from $\mathbf{P}'(x)$ to $x$ such that $f_1(y)\in y$ for every non-empty subset $y$ of $x$. Let $u$ be any set which does not belong to $x$. Then we define a mapping $f$ from $\mathbf{P}(x)$ to $x\cup\{u\}$ by setting

$$f(y)=\begin{cases} f_1(y) & \text{if } y\neq\emptyset \\ u & \text{if } y=\emptyset. \end{cases}$$

According to the Transfinite Recursion Theorem (see Exercise 99) there is a unique functional relation $G$ with domain $On$ such that for every ordinal number $\alpha$ we have $G(\alpha) = f(x \sim G^{\rightarrow}(\alpha))$.

(To get a feeling for the definition of $G$ we look at the determination of its values for the first few ordinals 0, 1, 2:

$G^{\rightarrow}(0) = G^{\rightarrow}(\emptyset) = \emptyset$; so $G(0) = f(x \sim \emptyset) = f(x)$.

$G^{\rightarrow}(1) = G^{\rightarrow}(\{0\}) = \{G(0)\} = \{f(x)\}$; so we have $G(1) = f(x \sim \{f(x)\}) = f(x_1)$ say;

$G^{\rightarrow}(2) = G^{\rightarrow}(\{0,1\}) = \{G(0), G(1)\} = \{f(x), f(x_1)\}$; so $G(2) = f(x \sim \{f(x), f(x_1)\}) = f(x_2)$.)

Let $Y = \{\alpha : (\alpha \in On) \wedge (G(\alpha) \neq u)\}$; we claim that $Y$ is a segment of $On$. To see this, suppose $\alpha \in Y$ and let $\beta$ be any element of $\alpha$. If $G(\beta) = f(x \sim G^{\rightarrow}(\beta)) = u$ we must have $x \sim G^{\rightarrow}(\beta) = \emptyset$, i.e. $G^{\rightarrow}(\beta) = x$. Since $\alpha$ is an ordinal and $\beta \in \alpha$ we have $\beta \subseteq \alpha$ and hence $G^{\rightarrow}(\alpha) \supseteq G^{\rightarrow}(\beta) = x$; it follows that $G(\alpha) = u$, which is a contradiction. Hence $G(\beta) \neq u$ and so $\beta \in Y$.

Thus $Y$ is a segment of $On$ and so is either $On$ itself or an ordinal number.

We claim that $G \mid Y$ is injective. So suppose that $\alpha$ and $\beta$ are elements of $Y$ with $\alpha \in \beta$. Then we have $G(\alpha) \in G^{\rightarrow}(\beta)$. But $G(\beta) = f(x \sim G^{\rightarrow}(\beta)) \in x \sim G^{\rightarrow}(\beta)$ and hence $G(\beta) \notin G^{\rightarrow}(\beta)$. Thus $G(\alpha) \neq G(\beta)$.

If $Y = On$ then $G$ is an injection from the proper class $On$ to the set $x$, which is impossible. So $Y$ is an ordinal number.

If $G^{\rightarrow}(Y) \neq x$ then $G(Y) = f(x \sim G^{\rightarrow}(Y)) \neq u$ and so $Y \in Y$, which contradicts the Axiom of Foundation. Hence $G$ is a bijection from the ordinal number $Y$ onto the set $x$, which can therefore be well-ordered.

112. (1) Let $X$ be a class such that $R \cap (X \times X)$ is a relation of total order on $X$.

Let $Y$ be any finite subclass of $X$. Let $a$ and $b$ be elements of $Y$. Since $a \in X$ and $b \in X$ we have either $(a, b) \in R$ or $(b, a) \in R$. So $R \cap (Y \times Y)$ is a relation of total order on $Y$.

Conversely, let $X$ be a class such that for every finite subclass $Y$ of $X$ the relation $R \cap (Y \times Y)$ is a total order on $Y$.

Let $a$ and $b$ be any elements of $X$. Take $Y = \{a, b\}$. Then either $(a, b) \in R$ or $(b, a) \in R$. So $R \cap (X \times X)$ is a relation of total order on $X$. Thus $P(X)$ is a property of finite character.

(2) Let $X$ be a class such that $u \notin X$.

Let $Y$ be any finite subclass of $X$. Then $u \notin Y$.

Conversely, let $X$ be a class such that for every finite subclass $Y$ of $X$ we have $u \notin Y$.

Then $u \notin X$ (for if $u \in X$ we have $u \in \{u\}$, which is a contradiction). So $P(X)$ is a property of finite character.

(3) Let $X$ be a class whose elements are pairwise disjoint sets.

Let $Y$ be a finite subclass of $X$. Then the elements of $Y$ are sets and if $a \in Y$ and $b \in Y$ we have $a \in X$ and $b \in X$ and so $a \cap b = \emptyset$.

Conversely, let $X$ be a class such that the elements of every finite subclass of $X$ are pairwise disjoint sets.

Let $a$ and $b$ be elements of $X$. Then $a$ and $b$ are elements of the finite subclass $\{a, b\}$ and so $a \cap b = \emptyset$. Hence $P(X)$ is a property of finite character.

113. (1) is not a theorem.

Let $P(X)$ be the property "$X \subseteq \omega$ and for all elements $m$, $n$ of $X$ we have $m \equiv n \bmod 2$." $P(X)$ is a property of finite character. If $x$ is the set of all even natural numbers and $y$ is the set of all odd natural numbers then $P(x)$ and $P(y)$ are true, but $P(x \cup y) = P(\omega)$ is not.

(2) is a theorem.

Let $t$ be any finite subset of $x \cap y$. Then $t$ is a finite subset of $x$ and since $P(x)$ is true it follows that $P(t)$ is true. So $P(x \cap y)$ is true.

114. Suppose the Hausdorff maximal principle holds.

Let $(x, R)$ be an inductively ordered set. According to the Hausdorff principle there is a $(\subseteq)$-maximal totally $(R)$-ordered subset $y$ of $x$. Since $(x, R)$ is inductively ordered, $y$ has an $R$-upper bound, say $u$. We claim that $u$ is an $R$-maximal element of $x$. If not, there is an element $v$ such that $(u, v) \in R$ and $v \neq u$. Clearly $v \notin y$ since $(t, u) \in R$ for all $t$ in $y$. Let $y_1 = y \cup \{v\}$; then $R \cap (y_1 \times y_1)$ is a total order on $y_1$ and $y_1 \supset y$. This contradicts the $(\subseteq)$-maximality of $y$.

So $u$ is an $R$-maximal element of $x$.

115. Suppose Zorn's Lemma holds.

 Let $x$ be a set, $P(X)$ a property of finite character. Let $Q = \{y : (y \subseteq x) \wedge P(y)\}$. We shall show that $Q$ is inductively ordered by the inclusion relation. So let $Q_0$ be a subset of $Q$ which is totally ordered by inclusion. Let $u = \bigcup Q_0$; then $u$ is an $(\subseteq)$-upper bound of $Q_0$ in $\mathbf{P}(x)$; we claim that $u \in Q$. So let $v$ be any finite subset of $u$, say $v = \{v_1, \ldots, v_n\}$. Then there are sets $q_1, \ldots, q_n$ in $Q_0$ such that $v_1 \in q_1, \ldots, v_n \in q_n$. Since $Q_0$ is totally ordered by inclusion one of the sets $q_1, \ldots, q_n$ includes all the others; say $q_1 \supseteq q_j$ $(j = 1, \ldots, n)$. Then $v$ is a finite subset of $q_1$ and since $q_1 \in Q$ it follows that $P(v)$ holds.

 Thus $P(u)$ holds, as required.

116. Suppose the Teichmüller-Tukey Lemma holds.

 Let $(x, R)$ be an ordered set. Let $P(X)$ be the property

$$\text{“}R \cap (X \times X) \text{ is a relation of total order on } X\text{”.}$$

 According to Exercise 112 $P(X)$ is a property of finite character. It follows that $Q = \{y : (y \in \mathbf{P}(x)) \wedge P(y)\}$ has an $(\subseteq)$-maximal element.

117. Suppose Zorn's Lemma holds.

 Let $(x_i)_{i \in I}$ be a family of non-empty sets.

 If $I = \emptyset$ then $\prod_{i \in I} x_i = \mathrm{Map}(\emptyset, \bigcup_{i \in I} x_i)$ is non-empty.

 So suppose $I \neq \emptyset$. Let

$$C = \{F : (F \text{ is a functional relation}) \wedge (\mathrm{Dom}\, F \subseteq I) \\ \wedge ((\forall i)((i \in \mathrm{Dom}\, F) \Longrightarrow (F(i) \in x_i)))\}.$$

 We shall show that $C$ is inductively ordered by the inclusion relation. So let $C_0$ be a subset of $C$ totally ordered by inclusion. Let $u = \bigcup C_0$; if we can show that $u \in C$ it will clearly be an $(\subseteq)$-upper bound for $C_0$ in $C$. Certainly $u$ is a set of ordered pairs, i.e. a relation, and $\mathrm{Dom}\, u = \bigcup_{g \in C_0} \mathrm{Dom}\, g \subseteq I$. To show that $u$ is a functional relation suppose that $(i, t_1) \in g_1$ and $(i, t_2) \in g_2$. Since $C_0$ is totally ordered by inclusion, either $g_1 \subseteq g_2$ or $g_2 \subseteq g_1$, say $g_1 \subseteq g_2$. Then $(i, t_1)$ and $(i, t_2)$ both belong to the functional relation $g_2$ and so $t_1 = t_2$. Thus $u$ is a functional relation. Let now $i$ be any element of Dom $u$. There is a relation $g$ in $C_0$ such that $i \in$ Dom $g$. Hence $u(i) = g(i) \in x_i$. So $u \in C$ and hence $C_0$ has an $(\subseteq)$-upper bound in $C$. According to Zorn's Lemma, $C$ has an $(\subseteq)$-maximal element, $v$ say. We claim that Dom $v = I$, and hence $v \in \prod_{i \in I} x_i$. Suppose, to the contrary, that Dom $v \neq I$ and let $j \in I \sim$ Dom $v$. If $a_j$ is any element of the non-empty set $x_j$ then we have $w = v \cup \{(j, a_j)\} \in C$ and $w \supset v$. This contradicts the maximal property of $u$.

So $\prod_{i \in I} x_i \neq \emptyset$, i.e. the Axiom of Choice holds.

118. Let $(x, R)$ be an ordered set.

Since the Axiom of Choice holds we deduce from Zermelo's Theorem that there is a relation of well-ordering on $x$ and hence (by Exercise 96) there is a bijection $f$ from some ordinal number $\alpha$ onto $x$. We define a mapping $g$ from $\alpha$ to $x$ by setting for each element $\beta$ of $\alpha$

$$g(\beta) = \begin{cases} f(\beta) & \text{if for all elements } \gamma \text{ of } \beta \text{ we have} \\ & \quad \text{either } (g(\gamma), f(\beta)) \in R \text{ or } (f(\beta), g(\gamma)) \in R \\ f(\emptyset) & \text{otherwise.} \end{cases}$$

We claim that the image $g^{\rightarrow}(\alpha)$ is a $(\subseteq)$-maximal totally ordered subset of $x$.

We notice first that $g(\emptyset) = f(\emptyset)$.

Then, if $g(\beta_1) = f(\beta_1)$ and $g(\beta_2) = f(\beta_2)$ are elements of $g^{\rightarrow}(\alpha)$ where $\beta_1 \in \beta_2$, we have either $(g(\beta_1), f(\beta_2)) \in R$ or else $(f(\beta_2), g(\beta_1)) \in R$, i.e. either $(g(\beta_1), g(\beta_2))$ or $(g(\beta_2), g(\beta_1))$ belongs to $R$. Thus $g^{\rightarrow}(\alpha)$ is a subset of $x$ totally ordered by $R$.

Suppose $f(\beta) \notin g^{\rightarrow}(\alpha)$. Then there must be an element $\gamma$ of $\beta$ for which either $(g(\gamma), f(\beta)) \notin R$ or $(f(\beta), g(\gamma)) \notin R$. Thus $g^{\rightarrow}(\alpha) \cup \{f(\beta)\}$ is not totally ordered by $R$. So $g^{\rightarrow}(\alpha)$ is maximal.

119. Let $t_0$ be any set in $T$. Let $T_0 = \{t : (t \in T) \wedge (t \supseteq t_0)\}$. We show that $T_0$ is inductively ordered by inclusion. So let $T_1$ be a subset of $T_0$ which is totally ordered by inclusion. Then $u = \bigcup T_1$ is an $(\subseteq)$-upper bound of $T_1$ in $\mathbf{P}(E)$; we have to show that $u \in T_0$.

Certainly $u \supseteq t_0$. So we have only to show that $u \in T$. To this end let $a$ and $b$ be any elements of $u$; then there are sets $s$ and

$t$ in $T_1$ such that $a \in s$ and $b \in t$. Since $T_1$ is totally ordered by inclusion we have either $s \subseteq t$ or $t \subseteq s$, say $s \subseteq t$. Then $a$ and $b$ both belong to $t$, which is totally ordered by $R \cap (t \times t)$; so either $(a, b) \in R$ or $(b, a) \in R$. Thus $u$ is totally ordered by $R \cap (u \times u)$. So $T_1$ has an $(\subseteq)$-upper bound in $T_0$.

Thus $T_0$ is inductively ordered by inclusion and hence has an $(\subseteq)$-maximal element, as required.

120. Let $P(X)$ be the property "$(X \subseteq \mathbf{R}) \wedge (X$ is $\mathbf{Q}$-free $)$".

Suppose $P(X)$ holds; if $Y$ is any finite subset of $X$ then certainly $P(Y)$ holds.

Conversely, suppose that $P(Y)$ holds for every finite subset $Y$ of $X$. Suppose $\sum_{x \in X} a_x x = 0$ where $(a_x)_{x \in X}$ is a family of rational numbers, only finitely many of which are nonzero. Then $Y = \{x : (x \in X) \wedge (a_x \neq 0)\}$ is finite and we have $\sum_{x \in Y} a_x x = 0$. Since $Y$ is free it follows that $a_x = 0$ for all $x$ in $Y$, hence for all $x \in X$. So $P(X)$ holds.

According to the Teichmüller-Tukey Lemma there is an $(\subseteq)$ -maximal $\mathbf{Q}$-free subset $B$ of $\mathbf{R}$. Suppose $B$ does not span $\mathbf{R}$ over $\mathbf{Q}$, i.e. $\mathbf{Q}B \neq \mathbf{R}$. Let $r$ be any real number not in $\mathbf{Q}B$. Then $B \cup \{r\}$ is free, contradicting the maximality of $B$. Thus $B$ is a Hamel basis.

# Chapter 24

# ANSWERS TO CHAPTER 11

121. As suggested, let $x = \{n : (n \in \omega) \wedge (\text{ every subset of } n \text{ is finite})\}$.

Then $\emptyset \in x$ since the only subset of $\emptyset$ is $\emptyset$ itself which is finite, being equipotent to the natural number $\emptyset$.

Suppose now that $k \in x$. To show that $k^+ \in x$ let $y$ be any subset of $k^+ = k \cup \{k\}$.

First, if $y \subseteq k$ then, since $k \in x$, we deduce that $y$ is finite. Next, if $y \nsubseteq k$, so that $k \in y$, it may happen that $y = k^+$. Since $k^+$ is a natural number it follows that $y$ is finite. Finally, suppose that $k \in y$ but $y \neq k^+$. Then there is an element $m$ of $k$ such that $m \notin y$. Let $z = (y \sim \{k\}) \cup \{m\}$, which is clearly equipotent to $y$; further, $z \subseteq k$ and hence is finite (since $k \in x$). Thus $y$ is finite.

Hence, in all cases, if $k \in x$ we have $k^+ \in x$.

Thus, by Exercise 106, $x = \omega$ and our assertion is established.

122. As suggested, let $x = \{n : (n \in \omega) \wedge (n \text{ is not equipotent to any proper subset of } n)\}$.

Clearly $\emptyset \in x$ since $\emptyset$ has no proper subset; again $\emptyset^+ = \{\emptyset\} \in x$ since its only proper subset is $\emptyset$ and $\{\emptyset\}$ is not equipotent to $\emptyset$.

Suppose now that some $k$ ($k \neq \emptyset$) is in $x$. To prove that $k^+ \in x$ we proceed by contradiction; so we suppose that there is a proper subset $y$ of $k^+ = k \cup \{k\}$ and a bijection $f$ of $k^+$ onto $y$.

First suppose that $k \notin y$. Then $y \subseteq k$ and $y_1 = y \sim \{f(k)\}$ is a proper subset of $k$. Now $f \mid k$ is a bijection of $k$ onto its proper subset $y_1$; but this is impossible since $k \in x$.

Next suppose that $k \in y$; then $y_2 = y \sim \{k\}$ is a proper subset of $k$.

If $f(k) = k$ then $f \mid k$ is a bijection of $k$ onto its proper subset $y_2$, which is impossible since $k \in x$.

If, on the other hand, $f(k) = r \neq k$ then of course $r \in y_2$; there must be an element $s$ of $k$ such that $f(s) = k$. Define a mapping $f_1$ from $k$ to $y_1$ by setting

$$f_1(t) = \begin{cases} f(t) & \text{if } t \neq s \\ r & \text{if } t = s. \end{cases}$$

Then $f_1$ is a bijection from $k$ onto its proper subset $y_2$, which is again a contradiction.

So if $k \in x$ then $k^+ \in x$ and hence, by Exercise 106, we have $x = \omega$.

123. The mapping $f$ from $\omega$ to $\omega$ given by $f(k) = k^+$ for all natural numbers $k$ is a bijection from $\omega$ onto $\omega \sim \{\emptyset\}$. (Use Exercises 103, 104 and 105.)

124. Suppose a finite set $x$ is equipotent to two distinct natural numbers $m$ and $n$.

Since $E \mid On$ is a relation of well-ordering on $On$ we have either $m \in n$ or $n \in m$, say $m \in n$. Then, since every element of an ordinal number is a subset, we have $m \subseteq n$ and hence (since $m \neq n$) $m$ is a proper subset of $n$.

Since $n$ is equipotent to $x$ and $x$ is equipotent to $m$ we deduce that $n$ is equipotent to its proper subset $m$. This contradicts the finiteness of $n$.

125. Since we are assuming that the Axiom of Choice holds we deduce from Zermelo's Theorem (Exercise 111) that there is a relation $R$ of well-ordering on $A$.

According to Exercise 85 either (1) there exists a unique isomorphism from $\omega$ onto $A$ or (2) there exists a unique isomorphism from $\omega$ onto an initial segment of $A$ or (3) there exists a unique isomorphism from $A$ onto an initial segment of $\omega$. Case (3) cannot occur for then we would have a bijection from the infinite set $A$ onto an initial segment of $\omega$, which is a natural number, and this is a contradiction. In cases (1) and (2) the image of $\omega$ under the unique isomorphism is the required denumerable subset.

126. (1) By Exercise 122 finite sets are not Dedekind-infinite. So Dedekind-infinite sets must be infinite.

(2) Conversely, let $A$ be an infinite set.

By Exercise 125 there is an injection $j$ of $\omega$ into $A$. Define a mapping $f$ from $A$ to $A$ by setting

$$f(x) = \begin{cases} x & \text{if } x \notin j^{\rightarrow}(\omega) \\ j(n^+) & \text{if } x = j(n) \text{ with } n \in \omega. \end{cases}$$

This mapping is injective but not surjective, since $j(\emptyset) \notin f^{\rightarrow}(A)$. So $A$ is equipotent to $f^{\rightarrow}(A)$, which is a proper subset of $A$, i.e. $A$ is Dedekind-infinite.

# Chapter 25

# ANSWERS TO CHAPTER 12

127. (a) $\emptyset$ is the only ordinal (indeed the only set) equinumerous with $\emptyset$. So Card $\emptyset = \emptyset$.

     (b) If $a$ is any set the mapping $f$ from $\{a\}$ to 1 given by $f(a) = \emptyset$ is a bijection. So $\{a\}$ is equinumerous with 1 but clearly not with any ordinal number $E$-less than 1 (there is only one, namely 0). Hence Card $\{a\} = 1$.

     (c) If $a$ and $b$ are distinct sets the mapping $f$ from $\{a, b\}$ to 2 given by $f(a) = 0$ and $f(b) = 1$ is a bijection. So $\{a, b\}$ is equinumerous with 2 but not with either of the ordinal numbers $E$-less than 2 (namely 0 and 1). So Card $\{a, b\} = 2$.

128. (1) Suppose Card $x$ = Card $y$.

     Since $x$ is equinumerous with Card $x$ and $y$ is equinumerous with Card $y$, it follows that $x$ is equinumerous with $y$.

     (2) Conversely, suppose $x$ is equinumerous with $y$.

     Then Card $x$ is equinumerous with $y$. Thus Card $x \in N(y)$. So (Card $y$, Card $x$) $\in E$. Similarly (Card $x$, Card $y$) $\in E$. So Card $x$ = Card $y$.

129. By Exercise 124 a finite set $x$ is equipotent to a unique natural number, $n$ say. Thus $n$ is the least (because it is the only) member of $N(x)$. So Card $x = n$.

     Conversely, if Card $x = n$ where $n$ is a natural number, then $x$ is equipotent to $n$ and hence is finite.

130. There are bijections $f$ and $g$ from $a$ to Card $a$ and $b$ to Card $b$ respectively.

If Card $a \leq$ Card $b$ then either (1) Card $a =$ Card $b$ in which case $a$ is equipotent to $b$ and so there is a bijection (which is of course an injection) from $a$ to $b$ or (2) Card $a \in$ Card $b$ and hence Card $a \subseteq$ Card $b$ since Card $b$ is an ordinal; then $g^{-1} \circ i \circ f$ is an injection from $a$ to $b$, where $i$ is the mapping of Card $a$ to Card $b$ given by $i(\xi) = \xi$ for every element $\xi$ of Card $a$.

Conversely, suppose there exists an injection $j$ from $a$ to $b$. Write $\alpha =$ Card $a$, $\beta =$ Card $b$. Let $x = j^{\rightarrow}(a)$. Define a relation $R$ on $b$ by setting

$$(s, t) \in R \iff (g(s), g(t)) \in E \mid \beta.$$

Since $E \mid \beta$ is a well-ordering on $\beta$ we see that $R$ is a well-ordering on $b$; clearly Ord $(b, R) = \beta$. Let Ord $(x, R \mid x) = \xi$. Then since Card $x$ is the $E$-least ordinal equinumerous with $x$ and $\xi$ is equinumerous with $x$ we have Card $x \leq \xi$. We shall show that $\xi \leq \beta$.

Since Ord $(b, R) = \beta$ there is a mapping $\phi$ from $b$ to $On$ which is strictly increasing with respect to the orders $R$ and $E$ on $b$ and $On$ respectively and is such that $\phi^{\rightarrow}(b) = \beta$. Similarly there is a mapping $\psi$ from $x$ to $On$, strictly increasing with respect to $R \mid x$ and $E$ such that $\psi^{\rightarrow}(x) = \xi$. Let $k$ be the natural injection of $x$ into $b$. Then $\psi$ and $\phi \circ k$ satisfy the conditions imposed on the mappings $f$ and $g$ respectively in Exercise 82. So for each element $s$ of $x$ we have $\psi(s) \leq \phi(s)$. Hence we have $\xi \subseteq \beta$.

It follows from this that $(\xi, \beta) \in E$. For, if not, we would have $\beta \in \xi$ and hence $\beta \subseteq \xi$; but then we would have $\beta = \xi$ and so $\beta \in \beta$ which contradicts the Axiom of Foundation.

Hence $\alpha =$ Card $a =$ Card $x \leq \xi \leq \beta$ as required.

131. Since there are injections from $a$ to $b$ and from $b$ to $a$ it follows from Exercise 127 that Card $a \leq$ Card $b$ and Card $b \leq$ Card $a$. Hence, since $\leq$ $(= E \mid Cn)$ is an order relation on $Cn$ we have Card $a =$ Card $b$ and hence, by Exercise 128, we see that $a$ is equipotent to $b$.

132. The mapping $i$ from $a$ to $\mathbf{P}(a)$ given by $i(t) = \{t\}$ for each element $t$ of $a$ is clearly an injection. Hence Card $a \leq$ Card $\mathbf{P}(a)$.

Suppose Card $a =$ Card $\mathbf{P}(a)$. Then there is a bijection $f$ from $a$ onto $\mathbf{P}(a)$. For each element $x$ of $a$ we have $f(x) \subseteq a$. Let

$$y = \{t : (t \in a) \wedge (t \notin f(t))\}.$$

Then $y \in \mathbf{P}(a)$. Since $f$ is surjective there is an element $x_1$ of $a$ such that $f(x_1) = y$. If $x_1 \in y$ then $x_1 \notin f(x_1) = y$ which is a contradiction. On the other hand, if $x_1 \notin y$ then $x_1 \in f(x_1) = y$ which is again a contradiction.

It follows that Card $a \neq$ Card $\mathbf{P}(a)$, as required.

133. Suppose $Cn$ were a set. Then by Axiom 5 (the Union Axiom) $X = \bigcup Cn$ would also be a set and hence, by Axiom 3 (the Power Set Axiom), $\mathbf{P}(X)$ would be a set. Then Card $(\mathbf{P}(X)) \subseteq X$ and so Card $\mathbf{P}(X) =$ Card(Card $\mathbf{P}(X)) \leq$ Card $X$. But by Cantor's Theorem we have Card $X <$ Card $\mathbf{P}(X)$. This contradiction shows that $Cn$ cannot be a set.

134. Consider the mapping $F$ from $\mathbf{P}(x)$ to Map$(x, 2)$ defined by setting $F(a) = \chi_a$ for all subsets $a$ of $x$. Let $G$ be the mapping from Map$(x, 2)$ to $\mathbf{P}(x)$ defined by setting $G(f) = f^{\leftarrow}(\{1\})$ for all mappings $f$ from $x$ to 2.

Then for each subset $a$ of $x$ we have

$$(G \circ F)(a) = G(\chi_a) = \chi_a^{\leftarrow}(\{1\}) = a,$$

i.e. $G \circ F = I_{\mathbf{P}(x)}$.

For each mapping $f$ from $x$ to 2 let $a_f = f^{\leftarrow}(\{1\})$. Then

$$f(t) = 1 \iff t \in a_f \iff \chi_{a_f}(t) = 1,$$

i.e. $f = \chi_{a_f}$. Thus we have

$$(F \circ G)(f) = F(a_f) = \chi_{a_f} = f,$$

i.e. $F \circ G = I_{\mathrm{Map}(x,2)}$. Hence $F$ and $G$ are bijections and so Card $\mathbf{P}(x) =$ Card (Map $(x, 2)$).

# Chapter 26

# ANSWERS TO CHAPTER 13

135. Let $f$ and $g$ be bijections from $\mathfrak{a}$ to $A$ and $\mathfrak{b}$ to $B$ respectively. Define the mapping $h$ from $(\mathfrak{a} \times \{0\}) \cup (\mathfrak{b} \times \{1\})$ to $A \cup B$ by setting

$$h(x) = \begin{cases} f(a) & \text{if } x = (a, 0) \\ g(b) & \text{if } x = (b, 1). \end{cases}$$

Then $h$ is a bijection and hence it follows that $\mathfrak{a} + \mathfrak{b} = \text{Card}\ ((\mathfrak{a} \times \{0\}) \cup (\mathfrak{b} \times \{1\})) = \text{Card}\ (A \cup B)$.

136. Let $f$ and $g$ be bijections from $\mathfrak{a}$ to $A$ and $\mathfrak{b}$ to $B$ respectively. Define the mapping $h$ from $\mathfrak{a} \times \mathfrak{b}$ to $A \times B$ by setting

$$h((a, b)) = (f(a), g(b))$$

for all elements $(a, b)$ of $\mathfrak{a} \times \mathfrak{b}$. Then $h$ is a bijection and hence $\mathfrak{a} \cdot \mathfrak{b} = \text{Card}(\mathfrak{a} \times \mathfrak{b}) = \text{Card}\ (A \times B)$.

137. (1) Let $A$ be a set equipotent to $\mathfrak{a}$; $\emptyset$ is equipotent to (actually equal to) the cardinal 0. Hence (using Exercise 135) we have

$$\mathfrak{a} + 0 = \text{Card}(A \cup \emptyset) = \text{Card}(A) = \mathfrak{a}.$$

$\{\emptyset\}$ is equipotent to (actually equal to) the cardinal 1. So we have $\mathfrak{a} \cdot 1 = \text{Card}\ (A \times \{\emptyset\})$. Now $A \times \{\emptyset\}$ is equipotent to $A$ under the bijection $f$ defined by setting $f((a, \emptyset)) = a$ for all elements $(a, \emptyset)$ of $A \times \{\emptyset\}$. Hence $\mathfrak{a} \cdot 1 = \text{Card}\ (A) = \mathfrak{a}$.

(2) Since $\mathfrak{a} \times \emptyset = \emptyset$ we have $\mathfrak{a} \cdot 0 = \text{Card}\ (\mathfrak{a} \times \emptyset) = \text{Card}\ \emptyset = 0$.

(3) Let $A$ and $B$ be disjoint sets equipotent to $\mathfrak{a}$ and $\mathfrak{b}$ respectively. Then we have

$$\mathfrak{a} + \mathfrak{b} = \text{Card}(A \cup B) = \text{Card}(B \cup A) = \mathfrak{b} + \mathfrak{a}.$$

Let $A$ and $B$ be sets equipotent to $\mathfrak{a}$ and $\mathfrak{b}$ respectively. Then $A \times B$ is equipotent to $B \times A$ under the mapping $f$ defined by setting $f((a, b)) = (b, a)$ for all elements $(a, b)$ of $A \times B$. So

$$\mathfrak{a} \cdot \mathfrak{b} = \text{Card}(A \times B) = \text{Card}(B \times A) = \mathfrak{b} \cdot \mathfrak{a}.$$

(4) Let $A$, $B$, $C$ be pairwise disjoint sets equipotent to $\mathfrak{a}$, $\mathfrak{b}$, $\mathfrak{c}$ respectively. Then $\mathfrak{a} + \mathfrak{b} = \text{Card } (A \cup B)$. So $A \cup B$ is equipotent to $\mathfrak{a} + \mathfrak{b}$. $C$ is disjoint from $A \cup B$ and equipotent to $\mathfrak{c}$. Hence we have $(\mathfrak{a} + \mathfrak{b}) + \mathfrak{c} = \text{Card } ((A \cup B) \cup C)$. Similarly $\mathfrak{a} + (\mathfrak{b} + \mathfrak{c}) = \text{Card } (A \cup (B \cup C))$. Since $(A \cup B) \cup C = A \cup (B \cup C)$ (as we noted in Chapter 2) we have $(\mathfrak{a} + \mathfrak{b}) + \mathfrak{c} = \mathfrak{a} + (\mathfrak{b} + \mathfrak{c})$.

Let $A$, $B$, $C$ be sets equipotent to $\mathfrak{a}$, $\mathfrak{b}$, $\mathfrak{c}$ respectively. Then $\mathfrak{a} \cdot \mathfrak{b} = \text{Card } (A \times B)$. So $A \times B$ is equipotent to $\mathfrak{a} \cdot \mathfrak{b}$. $C$ is equipotent to $\mathfrak{c}$. So $(\mathfrak{a} \cdot \mathfrak{b}) \cdot \mathfrak{c} = \text{Card}((A \times B) \times C)$. Similarly $\mathfrak{a} \cdot (\mathfrak{b} \cdot \mathfrak{c}) = \text{Card } (A \times (B \times C))$. But the mapping $h$ from $(A \times B) \times C$ to $A \times (B \times C)$ defined by setting $h(((a, b), c)) = (a, (b, c))$ for all elements $((a, b), c)$ of $(A \times B) \times C$ is clearly a bijection. So we have $\text{Card } ((A \times B) \times C) = \text{Card } (A \times (B \times C))$, i.e. $(\mathfrak{a} \cdot \mathfrak{b}) \cdot \mathfrak{c} = \mathfrak{a} \cdot (\mathfrak{b} \cdot \mathfrak{c})$.

(5) Let $A$ be equipotent to $\mathfrak{a}$; let $B$ and $C$ be disjoint sets equipotent to $\mathfrak{b}$ and $\mathfrak{c}$ respectively. Then $\mathfrak{b} + \mathfrak{c} = \text{Card } (B \cup C)$. So $B \cup C$ is equipotent to $\mathfrak{b} + \mathfrak{c}$. Now $A$ is equipotent to $\mathfrak{a}$; hence $\mathfrak{a} \cdot (\mathfrak{b} + \mathfrak{c}) = \text{Card } (A \times (B \cup C))$. Since $B$ and $C$ are disjoint so are $A \times B$ and $A \times C$ and these sets are equipotent to $\mathfrak{a} \times \mathfrak{b}$ and $\mathfrak{a} \times \mathfrak{c}$ respectively. Then

$$\begin{aligned} \mathfrak{a} \cdot \mathfrak{b} + \mathfrak{a} \cdot \mathfrak{c} &= \text{Card}((A \times B) \cup (A \times C)) \\ &= \text{Card}(A \times (B \cup C)) \\ &= \mathfrak{a} \cdot (\mathfrak{b} + \mathfrak{c}). \end{aligned}$$

138. Let $E = \mathfrak{a} + 1 = \mathfrak{b} + 1$.

Then $E$ has subsets $A$ and $B$ equipotent to $\mathfrak{a}$ and $\mathfrak{b}$ respectively and elements $x$, $y$ such that $x \notin A$, $y \notin B$ and $E = A \cup \{x\} = B \cup \{y\}$.

If $x = y$ then $A = B$ and so $\mathfrak{a} = \mathfrak{b}$.

If $x \neq y$ then $x \in B$ and $y \in A$; so $A = (A \cap B) \cup \{y\}$ and $B = (A \cap B) \cup \{x\}$. Hence $\mathfrak{a} =$Card$\,(A \cap B) + 1 = \mathfrak{b}$.

139. Let $n$ be a natural number.

Then $n + 1 = \text{Card}\,((n \times \{0\}) \cup (1 \times \{1\}))$.

Now $(n \times \{0\}) \cup (1 \times \{1\})$ is clearly equipotent to $n \cup \{n\}$ and hence $n + 1 = n^+$.

140. Let $f$ and $g$ be bijections from the cardinals $\mathfrak{a}$ and $\mathfrak{b}$ to the sets $A$ and $B$ respectively.

Let $F$ be the mapping from $\text{Map}\,(\mathfrak{b}, \mathfrak{a})$ to $\text{Map}\,(B, A)$ defined by setting

$$F(\varphi) = f \circ \varphi \circ g^{-1}$$

for all mappings $\varphi$ from $\mathfrak{b}$ to $\mathfrak{a}$. Then $F$ is easily shown to be a bijection and so $\mathfrak{a}^{\mathfrak{b}} = \text{Card Map}\,(\mathfrak{b}, \mathfrak{a}) = \text{Card Map}\,(B, A)$.

141. (1)Let $\mathfrak{a}$ be any cardinal.

Then $\text{Map}\,(0, \mathfrak{a}) = \text{Map}\,(\emptyset, \mathfrak{a}) = \{((\emptyset, \mathfrak{a}), \emptyset)\}$ which has cardinal 1. So $\mathfrak{a}^0 = 1$.

The mapping $F$ from $\text{Map}\,(1, \mathfrak{a}) = \text{Map}\,(\{\emptyset\}, \mathfrak{a})$ to $\mathfrak{a}$ defined by setting $F(f) = f(\emptyset)$ for each mapping $f$ from 1 to $\mathfrak{a}$ is clearly a bijection. So $\mathfrak{a}^1 = \text{Card Map}\,(1, \mathfrak{a}) = \text{Card}\ \mathfrak{a} = \mathfrak{a}$.

The only mapping from $\mathfrak{a}$ to $1 = \{\emptyset\}$ is the mapping $u$ given by $u(a) = \emptyset$ for all elements $a$ of $\mathfrak{a}$. So $\text{Map}\,(\mathfrak{a}, 1) = \{u\}$, which has cardinal 1. So $1^{\mathfrak{a}} = 1$.

The cardinals $2 \cdot \mathfrak{a}$ and $\mathfrak{a} + \mathfrak{a}$ are defined to be $\text{Card}(\{0, 1\} \times \mathfrak{a})$ and $\text{Card}(\mathfrak{a} \times \{0\} \cup \mathfrak{a} \times \{1\})$ respectively. We define a mapping $\varphi$ from $\{0, 1\} \times \mathfrak{a}$ to $\mathfrak{a} \times \{0\} \cup \mathfrak{a} \times \{1\}$ by setting $\varphi(i, a) = (a, i)$ for each $i$ in $\{0, 1\}$ and all $a$ in $\mathfrak{a}$. It is easily checked that $\varphi$ is a bijection and it follows at once that $2 \cdot \mathfrak{a} = \mathfrak{a} + \mathfrak{a}$.

By definition, $\mathfrak{a}^2 = \text{Card Map}\ (\{0, 1\}, \mathfrak{a})$ and $\mathfrak{a} \cdot \mathfrak{a} = \text{Card}\,(\mathfrak{a} \times \mathfrak{a})$. Define a mapping $\varphi$ from $\text{Map}\,(\{0, 1\}, \mathfrak{a})$ to $\mathfrak{a} \times \mathfrak{a}$ by setting $F(f) = (f(0), f(1))$ for each mapping $f$ from $\{0, 1\}$ to $\mathfrak{a}$. It is easily verified that $F$ is a bijection. Hence $\mathfrak{a}^2 = \mathfrak{a} \cdot \mathfrak{a}$.

If $\mathfrak{a} \neq 0$ then, according to Exercise 27, we have $\text{Map}\,(\mathfrak{a}, 0) = \text{Map}\,(\mathfrak{a}, \emptyset) = \emptyset$. So $0^{\mathfrak{a}} = 0$.

(2) Since $\mathfrak{a} = \text{Card}(\mathfrak{a}) \leq \mathfrak{b} = \text{Card}(\mathfrak{b})$ it follows from Exercise 130 that there is an injection $\iota$ from $\mathfrak{a}$ to $\mathfrak{b}$.

The mappings $\iota_1$ from $\mathfrak{a} \times \{0\} \cup \mathfrak{c} \times \{1\}$ to $\mathfrak{b} \times \{0\} \cup \mathfrak{c} \times \{1\}$ given by

$$\iota_1(x) = \begin{cases} (\iota(a), 0) & \text{if } x = (a, 0) \text{ with } a \in \mathfrak{a}, \\ (c, 1) & \text{if } x = (c, 1) \text{ with } c \in \mathfrak{c} \end{cases}$$

and the mapping $\iota_2$ from $\mathfrak{a} \times \mathfrak{c}$ to $\mathfrak{b} \times \mathfrak{c}$ given by

$$\iota_2(a, c) = (\iota(a), c) \text{ for all } (a, c) \text{ in } \mathfrak{a} \times \mathfrak{c}$$

are easily seen to be injections. Hence we have $\mathfrak{a} + \mathfrak{c} = \text{Card}(\mathfrak{a} \times \{0\} \cup \mathfrak{c} \times \{1\}) \leq \text{Card}(\mathfrak{b} \times \{0\} \cup \mathfrak{c} \times \{1\}) = \mathfrak{b} + \mathfrak{c}$ and $\mathfrak{a} \cdot \mathfrak{c} = \text{Card}(\mathfrak{a} \times \mathfrak{c}) \leq \text{Card}(\mathfrak{b} \times \mathfrak{c}) = \mathfrak{b} \cdot \mathfrak{c}$.

142. (1) Let $A$ be a set equipotent to $\mathfrak{a}$; let $B$ and $C$ be disjoint sets equipotent to $\mathfrak{b}$ and $\mathfrak{c}$ respectively. Then $B \cup C$ is equipotent to $\mathfrak{b} + \mathfrak{c}$ and we have $\mathfrak{a}^{\mathfrak{b}+\mathfrak{c}} = \text{Card Map}\,(B \cup C, A)$, while $\mathfrak{a}^{\mathfrak{b}} \cdot \mathfrak{a}^{\mathfrak{c}} = \text{Card}\,(\text{Map}\,(B, A) \times \text{Map}\,(C, A))$.

Let $\iota_B$, $\iota_C$ respectively be the natural injections of $B$, $C$ into $B \cup C$. We then define the mapping $F$ from $\text{Map}\,(B \cup C, A)$ to $\text{Map}\,(B, A) \times \text{Map}\,(C, A)$ by setting

$$F(\varphi) = (\varphi \circ \iota_B, \varphi \circ \iota_C)$$

for all mappings $\varphi$ from $B \cup C$ to $A$. It is easily shown that $F$ is a bijection.

Hence $\mathfrak{a}^{\mathfrak{b}+\mathfrak{c}} = \mathfrak{a}^{\mathfrak{b}} \cdot \mathfrak{a}^{\mathfrak{c}}$.

(2) Let $A$, $B$, $C$ be sets equipotent to $\mathfrak{a}$, $\mathfrak{b}$, $\mathfrak{c}$ respectively. Then we have $\mathfrak{a}^{\mathfrak{b} \cdot \mathfrak{c}} = \text{Card Map}\,(B \times C, A)$ and on the other hand $(\mathfrak{a}^{\mathfrak{b}})^{\mathfrak{c}} = \text{Card Map}\,(C, \text{Map}\,(B, A))$.

By Exercise 38 there is a bijection from $\text{Map}\,(C, \text{Map}\,(B, A))$ to $\text{Map}\,(B \times C, A)$. So we have $\mathfrak{a}^{\mathfrak{b} \cdot \mathfrak{c}} = (\mathfrak{a}^{\mathfrak{b}})^{\mathfrak{c}}$.

(3) Again let $A$, $B$, $C$ be sets equipotent to $\mathfrak{a}$, $\mathfrak{b}$, $\mathfrak{c}$ respectively. Then we have $(\mathfrak{a} \cdot \mathfrak{b})^{\mathfrak{c}} = \text{Card Map}\,(C, A \times B)$ and also $\mathfrak{a}^{\mathfrak{c}} \cdot \mathfrak{b}^{\mathfrak{c}} = \text{Card}\,(\text{Map}\,(C, A) \times \text{Map}\,(C, B))$.

According to Exercise 39 there is a bijection from $\text{Map}\,(C, A \times B)$ to $\text{Map}\,(C, A) \times \text{Map}\,(C, B)$. So we have $(\mathfrak{a} \cdot \mathfrak{b})^{\mathfrak{c}} = \mathfrak{a}^{\mathfrak{c}} \cdot \mathfrak{b}^{\mathfrak{c}}$.

143. Let $A$ be a set equipotent to $\mathfrak{a}$, $(B_i)_{i \in I}$ a family of pairwise disjoint sets such that for each index $i$ in $I$ we have $B_i$ equipotent to $\mathfrak{b}_i$.

 Then $\mathfrak{a}^{\sum \mathfrak{b}_i}$ is equipotent to Map $(\bigcup B_i, A)$ and $\prod \mathfrak{a}^{\mathfrak{b}_i}$ is equipotent to $\prod$ Map $(B_i, A)$. Define a mapping $\varphi$ from Map $(\bigcup B_i, A)$ to $\prod$ Map $(B_i, A)$ by setting for each mapping $f$ from $\bigcup B_i$ to $A$

$$(\varphi(f))(i) = f \mid B_i \text{ for all indices } i \text{ in} I.$$

 Then $\varphi$ is a bijection and the result follows.

144. This is an immediate consequence of Exercise 131.

145. Let $\alpha$ be an ordinal equipotent to $\omega$.

 If $\alpha <_E \omega$ then $\alpha \in \omega$ and so $\alpha$ is a natural number, hence finite. This is a contradiction since $\omega$ is not finite (see Exercise 123).

 Thus $\omega$ is the $E$-least ordinal equipotent to $\omega$, i.e. $\omega =$ Card $\omega$.

146. (1) Define the mapping $h$ from $(\aleph_0 \times \{0\}) \cup (n \times \{1\})$ to $\aleph_0$ by setting

$$h(x) = \begin{cases} a + n & \text{if } x = (a, 0) \text{ with } a \in \aleph_0 \\ b & \text{if } x = (b, 1) \text{ with } b \in n. \end{cases}$$

 Then $h$ is a bijection and so we have $\aleph_0 + n = \text{Card}((\aleph_0 \times \{0\}) \cup (n \times \{1\})) = \text{Card } \aleph_0 = \aleph_0$.

 (2) Let $n$ be a non-zero natural number. Then $n$ is the successor of a natural number, $m$ say, and we have $n = \{0, 1, 2, \ldots, m\}$. Then $n \cdot \aleph_0$ is equipotent to $\{0, 1, 2, \ldots, m\} \times \aleph_0$. The mapping $F$ from this product to $\aleph_0$ given by $F(k, l) = k + l \cdot n$ for all $k$ in $\{0, 1, 2, \ldots, m\}$ and all natural numbers $l$ is a bijection.

 To prove this requires a little well-known elementary number theory which we have not formally developed. Namely, to show that $F$ is injective suppose we have $F(k_1, l_1) = F(k_2, l_2)$, i.e. $k_1 + l_1 n = k_2 + l_2 n$. Then $k_1 - k_2 = (l_2 - l_1)n$ is divisible by $n$; but $k_1$ and $k_2$ are less than $n$. Hence $k_1 - k_2$ must be 0, i.e. $k_1 = k_2$ and hence $l_1 = l_2$. To show that $F$ is surjective we make use of Euclid's algorithm, which shows that every natural number can be expressed in the form $ln + k$ with $k < n$.

147. Since $\mathfrak{a}$ is infinite we have $0 < \mathfrak{a}$ and $3 < \mathfrak{a}$. It follows, using part (2) of Exercise 141, that

$$\mathfrak{a} = \mathfrak{a} + 0 \leq \mathfrak{a} + \mathfrak{a} = 2 \cdot \mathfrak{a} \leq 3 \cdot \mathfrak{a} \leq \mathfrak{a} \cdot \mathfrak{a} = \mathfrak{a}^2 = \mathfrak{a}.$$

Hence $\mathfrak{a} = 2 \cdot \mathfrak{a} = 3 \cdot \mathfrak{a}$.

148. (1) $\aleph_0 + \aleph_0 = \mathrm{Card}\,(\aleph_0 \times \{1\}) \cup (\aleph_0 \times \{2\})$. The mapping $F$ from $(\aleph_0 \times \{1\}) \cup (\aleph_0 \times \{2\})$ to $\aleph_0$ given by

$$F(n, i) = \begin{cases} 2 \cdot n & \text{if } i = 0 \\ 2 \cdot n + 1 & \text{if } i = 1. \end{cases}$$

is a bijection. (Again we are using a number-theoretic result we have not formally proved, namely that every natural number is either odd or even.)

(2) Number the rows of an array by the natural numbers, starting with 0 at the top; number the columns by the natural numbers, starting with 0 at the left. Consider the array

$$\begin{array}{ccccc} 0 & 2 & 5 & 9 & 14 \\ 1 & 4 & 8 & 13 & \ldots \\ 3 & 7 & 12 & \ldots & \ldots \\ 6 & 11 & \ldots & \ldots & \ldots \\ 10 & \ldots & \ldots & \ldots & \ldots \end{array}$$

Examine first the positions $(0, n)$ in the leftmost column; 0 is placed in the $(0,0)$-th position. Now let $n$ be a non-zero natural number; before filling the $(0, n)$-th position we fill the $1+2+\cdots+n$ positions in the first $n$ lower-left-to-upper-right diagonals with the natural numbers starting with 0. Since it is well-known (though again this is not something we have developed formally) that $1+2+\cdots+n = \frac{1}{2}n(n+1)$ we have used the natural numbers from 0 to $\frac{1}{2}n(n+1)-1$ to fill the positions on the first $n$ diagonals; hence the $(0, n)$-th position is filled with the natural number $\frac{1}{2}n(n+1)$. Thus in defining a mapping $f$ from $\aleph_0 \times \aleph_0$ to $\aleph_0$ we begin by setting $f(0, n) = \frac{1}{2}n(n+1)$ for all natural numbers $n$.

Suppose now that $i > 0$. Then the $(i, n)$-th position lies on the diagonal with lower end position $(0, n+i)$ and is the $i$-th position on that diagonal (starting the count at 0). Hence the $(i, n)$-th position in the array is filled with $f(0, n+i) + i$.

So we define the mapping $f$ by setting

$$f(i,n) = \begin{cases} \frac{1}{2}n(n+1) & \text{if } i = 0 \\ f(0, n+i) + i & \text{if } i \neq 0. \end{cases}$$

It is not hard to show that $f$ is bijective. Hence we have $\aleph_0 \cdot \aleph_0 = \operatorname{Card}(\aleph_0 \times \aleph_0) = \operatorname{Card} \aleph_0 = \aleph_0$.

149. Let $E$ be the set of ordered pairs $(A, f)$ such that $A$ is a subset of $\mathfrak{a}$ and $f$ is a bijection from $A$ to $A \times A$.

$E$ is non-empty since (because it is infinite) $\mathfrak{a}$ has a subset $A$ equipotent to $\aleph_0$ and (because $\aleph_0 \cdot \aleph_0 = \aleph_0$) there is a bijection from $A$ to $A \times A$.

Let $R$ be the order relation on $E$ defined by setting

$$((A, f), (B, g)) \in R \iff ((A \subseteq B) \wedge (g \,|\, A = f)).$$

We claim that $E$ is inductively ordered by $R$.

So let $E'$ be a subset of $E$ totally ordered by $R$. Let $X$ be the union of the set of all sets which occur as first coordinates of ordered pairs in $E'$. Define a mapping $h$ from $X$ to $X \times X$ as follows: for each element $x$ of $X$ there is a pair $(A, f)$ in $E'$ such that $x \in A$; set $h(x) = f(x)$. (Note that $h(x)$ depends only on $x$, for, if there is another pair $(B, g)$ in $E'$ such that $x \in B$ then, since $E'$ is totally ordered by $R$, we have either $((A, f), (B, g)) \in R$ or $((B, g), (A, f)) \in R$, whence $f(x) = g(x)$.)

To show that $h$ is an injection, suppose $x_1$, $x_2$ are elements of $X$ such that $h(x_1) = h(x_2)$. If $x_1 \in A_1$ where $(A_1, f_1) \in E'$ and similarly $x_2 \in A_2$ where $(A_2, f_2) \in E'$ then we have either $A_1 \subseteq A_2$ and $f_2 \,|\, A_1 = f_1$ or else $A_2 \subseteq A_1$ and $f_1 \,|\, A_2 = f_2$. In the first case we have $h(x_1) = f_1(x_1) = f_2(x_1)$ and $h(x_2) = f_2(x_2)$ whence $f_2(x_1) = f_2(x_2)$ and hence $x_1 = x_2$ since $f_2$ is injective. In the second case we have similarly $h(x_2) = f_2(x_2) = f_1(x_2)$ and $h(x_1) = f_1(x_1)$, whence $f_1(x_1) = f_1(x_2)$ and so again $x_1 = x_2$. Thus $h$ is injective.

To show that $h$ is surjective let $(x, y)$ be any element of $X \times X$. Then $x \in X$ and $y \in X$ and so there are pairs $(A, f)$, $(B, g)$ in $E'$ such that $x \in A$ and $y \in B$. Since $E'$ is totally ordered by $R$ we have either $((A, f), (B, g)) \in R$ or $((B, g), (A, f)) \in R$. In the first case we have $A \subseteq B$ and hence $(x, y) \in B \times B$; since $g$ is surjective there is an element $b$ of $B$ such that $h(b) = g(b) = (x, y)$. Similarly in the second case there is an element $a$ of $A$ such that $h(a) = f(a) = (x, y)$. Thus $h$ is surjective.

This discussion shows that the ordered pair $(X, h)$ belongs to $E$; it is clearly an $R$-upper bound for $E'$. So $E$ is inductively ordered by $R$ and hence, by Zorn's Lemma, $E$ has an $R$-maximal element $(M, k)$.

$M$ is clearly infinite. We claim that in fact $\operatorname{Card} M = \mathfrak{a}$. Since $k$ is a bijection from $M$ to $M \times M$ it will follow that $\mathfrak{a}^2 = \mathfrak{a}$.

Suppose, to the contrary, that $\operatorname{Card} M < \mathfrak{a}$.

Since $M \times M$ is equipotent to $M$, we have $(\operatorname{Card} M)^2 = \operatorname{Card} M$ and so, by Exercise 147, $2 \cdot \operatorname{Card} M = \operatorname{Card} M$. Now if we had $\operatorname{Card}(\mathfrak{a} \sim M) \leq \operatorname{Card} M$ it would follow that $\mathfrak{a} = \operatorname{Card} M + \operatorname{Card}(\mathfrak{a} \sim M) \leq \operatorname{Card} M + \operatorname{Card} M = 2 \cdot \operatorname{Card} M = \operatorname{Card} M$, which is a contradiction. It follows that there is a subset $S$ of $\mathfrak{a} \sim M$ which is equipotent to $M$.

Since $(\operatorname{Card} S)^2 = \operatorname{Card} S$ we have $3 \cdot \operatorname{Card} S = \operatorname{Card} S$ (by Exercise 147). Now

$$(M \cup S) \times (M \cup S) = (M \times M) \cup (M \times S) \cup (S \times M) \cup (S \times S).$$

This union is disjoint; so we have

$$\begin{aligned} &\operatorname{Card}((M \times S) \cup (S \times M) \cup (S \times S)) \\ &\quad = (\operatorname{Card} M) \cdot (\operatorname{Card} E) + (\operatorname{Card} E) \cdot (\operatorname{Card} M) + \\ &\qquad\qquad + (\operatorname{Card} E) \cdot (\operatorname{Card} E) \\ &\quad = 3 \cdot (\operatorname{Card} S)^2 = 3 \cdot \operatorname{Card} S = \operatorname{Card} S. \end{aligned}$$

So there is a bijection $k'$ from $S$ to $(M \times S) \cup (S \times M) \cup (S \times S)$. Thus we have a bijection $\varphi$ from $M \cup S$ to $(M \cup S) \times (M \cup S)$ defined by setting

$$\varphi(x) = \begin{cases} k(x) & \text{if } x \in M \\ k'(x) & \text{if } x \in S. \end{cases}$$

So $(M \cup S, \varphi)$ is an element of $E$ which is $R$-greater than the $R$-maximal element $(M, k)$. This, of course, is impossible; so we deduce that $\operatorname{Card} M = \mathfrak{a}$ as asserted and so that $\mathfrak{a}^2 = \mathfrak{a}$.

150. Let $A$ be any non-empty subset of $X$.

If $A \cap (X_1 \times \{1\})$ is non-empty then $\pi_1^{\rightarrow}(A \cap (X_1 \times \{1\}))$ is a non-empty subset of $X_1$ and hence has an $R_1$-least member, $x_1$ say. Then $(x_1, 1)$ is the $R$-least element of $A$. If $A \cap (X_1 \times \{1\}) = \emptyset$ then $A \cap (X_2 \times \{2\})$ must be non-empty and so $\pi_1^{\rightarrow}(A \cap (X_2 \times \{2\})$ is a non-empty subset of $X_2$ and so has an $R_2$-least element, say $x_2$. Then $(x_2, 2)$ is the $R$-least element of $A$.

151. Let $A$ be a non-empty subset of $X_1 \times X_2$.

Since $A$ is non-empty, $\pi_2^{\rightarrow}(A)$ is non-empty and so has an $R_2$-least element, say $a_2$. The set $\pi_1^{\rightarrow}(\pi_2^{\leftarrow}(\{a_2\}))$ is then a non-empty subset of $X_1$ and hence has an $R_1$-least member, $a_1$ say. Then $(a_1, a_2)$ is the $S$-least element of $A$.

152. Suppose $f_1$ and $f_2$ are $(R_1, R_1')$- and $(R_2, R_2')$-isomorphisms from $X_1$ to $X_1'$ and $X_2$ to $X_2'$ respectively.

(1) Let $f$ be the mapping from $X_1 \oplus X_2$ to $X_1' \oplus X_2'$ defined by setting

$$f((x_1, 1)) = (f_1(x_1), 1) \text{ for all elements } x_1 \text{ of } X_1$$

and

$$f((x_2, 2)) = (f_2(x_2), 2) \text{ for all elements } x_2 \text{ of } X_2.$$

Then $f$ is an $(R_1 \oplus R_2, R_1' \oplus R_2')$-isomorphism from $X_1 \oplus X_2$ to $X_1' \oplus X_2'$.

(2) Let $g$ be the mapping from $X_1 \times X_2$ to $X_1' \times X_2'$ defined by setting

$$g(x_1, x_2) = (f_1(x_1), f_2(x_2)) \text{ for all elements } (x_1, x_2) \text{ of } X_1 \times X_2.$$

Then $g$ is an $(R_1 \otimes R_2, R_1' \otimes R_2')$-isomorphism from $X_1 \times X_2$ to $X_1' \times X_2'$.

153. (1) The mapping $f$ from $\alpha$ to $(\alpha \times \{1\}) \cup (\emptyset \times \{2\}) = \alpha \times \{1\}$ given by setting $f(a) = (a, 1)$ for all elements $a$ of $\alpha$ is easily verified to be an $(E \mid \alpha, E \mid \alpha \oplus E \mid \emptyset)$-isomorphism from $\alpha$ to $\alpha \oplus \emptyset$. Hence $\alpha = \text{Ord}\,(\alpha, E \mid \alpha) = \text{Ord}\,(\alpha \oplus \emptyset, E \mid \alpha \oplus E \mid \emptyset) = \alpha + 0$.

Similarly the mapping $g$ from $\alpha$ to $(\emptyset \times \{1\}) \cup (\alpha \times \{2\}) = \alpha \times \{2\}$ given by $g(a) = (a, 2)$ for all elements $a$ of $\alpha$ is easily seen to be an $(E \,|\, \alpha, E \,|\emptyset \oplus E \,|\, \alpha)$-isomorphism from $\alpha$ to $\emptyset \oplus \alpha$. Hence $\alpha = \mathrm{Ord}\,(\alpha, E \,|\, \alpha)) = \mathrm{Ord}\,(\emptyset \oplus \alpha, E \,|\, \emptyset \oplus E \,|\, \alpha) = 0 + \alpha$.

The mapping $h$ from $\alpha$ to $\alpha \times 1$ given by $h(a) = (a, 0)$ for all elements $a$ of $\alpha$ is an $(E \,|\, \alpha, E \,|\, \alpha \otimes E \,|\, \emptyset)$-isomorphism from $\alpha$ to $\alpha \times 1$. So $\alpha = \mathrm{Ord}\,(\alpha, E \,|\, \alpha) = \mathrm{Ord}\,(\alpha \times 1, E \,|\, \alpha \otimes E \,|\, 1) = \alpha \cdot 1$.

Similarly the mapping $k$ from $\alpha$ to $1 \times \alpha$ given by $k(a) = (0, a)$ for all elements $a$ of $\alpha$ is an $(E \,|\, \alpha, E \,|\, 1 \otimes E \,|\, \alpha)$-isomorphism from $\alpha$ to $1 \times \alpha$. So $\alpha = \mathrm{Ord}\,(\alpha, E \,|\, \alpha) = \mathrm{Ord}\,(1 \times \alpha, E \,|\, 1 \otimes E \,|\, \alpha) = 1 \cdot \alpha$.

(2) Since $\emptyset \times \alpha = \alpha \times \emptyset = \emptyset$ we have $0 \cdot \alpha = \mathrm{Ord}\,(\emptyset \times \alpha) = 0$ and $\alpha \cdot 0 = \mathrm{Ord}\,(\alpha \times \emptyset) = 0$.

(3) $\alpha + (\beta + \gamma)$ is isomorphic (with respect to the obvious order relations) to $\alpha \oplus (\beta + \gamma)$ and hence to

$$\begin{aligned} &\alpha \oplus (\beta \oplus \gamma) \\ &\quad = (\alpha \times \{1\}) \cup ((\beta \oplus \gamma) \times \{2\}) \\ &\quad = (\alpha \times \{1\}) \cup (((\beta \times \{1\}) \cup (\gamma \times \{2\})) \times \{2\}) \\ &\quad = (\alpha \times \{1\}) \cup (((\beta \times \{1\}) \times \{2\}) \cup ((\gamma \times \{2\}) \times \{2\})). \end{aligned}$$

Similarly $(\alpha + \beta) + \gamma$ is isomorphic to $(\alpha + \beta) \oplus \gamma$ and hence to

$$\begin{aligned} &(\alpha \oplus \beta) \oplus \gamma \\ &\quad = ((\alpha \oplus \beta) \times \{1\}) \cup (\gamma \times \{2\}) \\ &\quad = (((\alpha \times \{1\}) \cup (\beta \times \{2\})) \times \{1\}) \cup (\gamma \times \{2\}) \\ &\quad = (((\alpha \times \{1\}) \times \{1\}) \cup ((\beta \times \{2\}) \times \{1\})) \cup (\gamma \times \{2\}). \end{aligned}$$

In both cases the unions are disjoint. We define a mapping $f$ from the first union to the second by setting

$$f(x) = \begin{cases} ((a,1),1) & \text{if } x = (a,1) \text{ with } a \text{ in } \alpha \\ ((b,1),2) & \text{if } x = ((b,2),1) \text{ with } b \text{ in } \beta \\ (c,2) & \text{if } x = ((c,2),2) \text{ with } c \text{ in } \gamma. \end{cases}$$

Then $f$ is an isomorphism and the desired result follows.
By definition we have

$$\alpha \cdot (\beta \cdot \gamma) = \mathrm{Ord}\,(\alpha \times (\beta \times \gamma), E \,|\, \alpha \otimes (E \,|\, \beta \otimes E \,|\, \gamma))$$

and

$$(\alpha \cdot \beta) \cdot \gamma = \mathrm{Ord}\,((\alpha \times \beta) \times \gamma, (E \,|\, \alpha \otimes E \,|\, \beta) \otimes E \,|\, \gamma).$$

The mapping $f$ from $\alpha \times (\beta \times \gamma)$ to $(\alpha \times \beta) \times \gamma$ defined by setting $f(a,(b,c)) = ((a,b),c)$ for all elements $a$ of $\alpha$, $b$ of $\beta$, $c$ of $\gamma$ is easily seen to be an isomorphism relative to the relevant order relations. Hence, using Exercise 97, we see that $\alpha \cdot (\beta \cdot \gamma) = (\alpha \cdot \beta) \cdot \gamma$.

(4) $\alpha \cdot (\beta + \gamma)$ is isomorphic (with respect to the obvious order relations) to $\alpha \times (\beta \oplus \gamma)$ and hence to
$\alpha \times ((\beta \times \{1\}) \cup (\gamma \times \{2\})) = (\alpha \times (\beta \times \{1\})) \cup (\alpha \times (\gamma \times \{2\}))$.
On the other hand, $\alpha \cdot \beta + \alpha \cdot \gamma$ is isomorphic to $(\alpha \times \beta) \oplus (\alpha \times \gamma) = ((\alpha \times \beta) \times \{1\}) \cup ((\alpha \times \gamma) \times \{2\})$.

In both cases the unions are disjoint. We define a mapping $f$ from the first union to the second by setting

$$f(x) = \begin{cases} ((a,b),1) & \text{if } x = (a,(b,1)) \text{ with } a \in \alpha \text{ and } b \in \beta \\ ((a,c),2) & \text{if } x = (a,(c,2)) \text{ with } a \in \alpha \text{ and } c \in \gamma. \end{cases}$$

Then $f$ is an isomorphism and the desired result follows.

154. (1) $1 \oplus \omega = (1 \times \{1\}) \cup (\omega \times \{2\})$.

Define a mapping $f$ from $1 \oplus \omega$ to $\omega$ by setting

$$f((0,1)) = 0$$

and

$$f((n,2)) = n^+ \text{ for every natural number } n.$$

Then $f$ is an $(E \,|\, 1 \oplus E \,|\, \omega, E \,|\, \omega)$-isomorphism from $1 \oplus \omega$ to $\omega$ and hence $1 + \omega = \mathrm{Ord}\,(1 \oplus \omega) = \mathrm{Ord}\,\omega = \omega$.

(2) Suppose we did have $\omega + 1 = \omega$.

Then there would be an isomorphism $f$ from $\omega$ to $\omega \oplus 1$. Suppose $(0,2)$ in $1 \times \{2\}$ is the image under $f$ of the natural number $k$. Then $f(k^+)$ would be an element of $\omega \oplus 1$ which is $(E \,|\, \omega \oplus E \,|\, 1)$-greater than $(0,2)$; but there is no such element.

From this contradiction we deduce that $\omega + 1 \neq \omega$.

155. (1) Consider the mapping $f$ from $2 \times \omega$ to $\omega$ defined by setting for all natural number $k$

$$f((i,k)) = \begin{cases} 2 \cdot k & \text{if } i = 0 \\ 2 \cdot k + 1 & \text{if } i = 1 \end{cases}$$

Then $f$ is an $(E\,|\,2 \otimes E\,|\,\omega, E\,|\,\omega)$-isomorphism from $2 \times \omega$ to $\omega$ and hence $2 \cdot \omega = \mathrm{Ord}\,(2 \times \omega) = \mathrm{Ord}\,\omega = \omega$.

(2) Consider the mapping $f$ from $\omega \times 2$ to $\omega \oplus \omega$ defined by setting

$$f(x) = \begin{cases} (k, 1) & \text{if } x = (k, 0) \\ (k, 2) & \text{if } x = (k, 1). \end{cases}$$

Then $f$ is an $(E\,|\,\omega \otimes E\,|\,2, E\,|\,\omega \oplus E\,|\,\omega)$-isomorphism and hence we have $\omega \cdot 2 = \mathrm{Ord}\,(\omega \times 2) = \mathrm{Ord}\,(\omega \oplus \omega) = \omega + \omega$.

Let $R = E\,|\,\omega \oplus E\,|\,\omega$. If $\omega + \omega = \omega$ then there would be an $(E\,|\,\omega, R)$-isomorphism $f$ from $\omega$ to $\omega \oplus \omega = (\omega \times \{1\}) \cup (\omega \times \{2\})$. Hence there would be a natural number $k$ such that $f(k) = (0, 2)$. Clearly $k$ must be non-zero since $0$ is the $(E\,|\,\omega)$-first element of $\omega$ but $(0, 2)$ is not the $R$-first element of $\omega \oplus \omega$ (being preceded by all the elements of $\omega \times \{1\}$). Hence $k$ is the successor of a natural number: say $k = n^+$. Since $n < k$ we must have $f(n) <_R f(k) = (0, 2)$ and hence $f(n) = (l, 1)$ for some natural number $l$. There is no natural number $m$ such that $n < m$ and $m < n^+ = k$. So there can be no element $x$ of $\omega \oplus \omega$ such that $(l, 1) = f(n) <_R x$ and $x <_R f(n^+) = f(k) = (0, 2)$. But this is not the case since (for example) $(l^+, 1)$ satisfies these conditions. Hence we have $\omega + \omega \neq \omega$.

# Index